Wireless Networks

Series editor

Xuemin Sherman Shen
University of Waterloo, Waterloo, ON, Canada

More information about this series at http://www.springer.com/series/14180

Haipeng Yao • Chunxiao Jiang • Yi Qian

Developing Networks using Artificial Intelligence

Springer

Haipeng Yao
Sch of Info & Communication Engg
Beijing University of Posts and Telecomm
China, Beijing, China

Chunxiao Jiang
Tsinghua Space Center
Tsinghua University
Beijing, Beijing, China

Yi Qian
Dept of Electrical & Computer Engg
University of Nebraska-Lincoln
Omaha, NE, USA

ISSN 2366-1186 ISSN 2366-1445 (electronic)
Wireless Networks
ISBN 978-3-030-15027-3 ISBN 978-3-030-15028-0 (eBook)
https://doi.org/10.1007/978-3-030-15028-0

Library of Congress Control Number: 2019935167

This Springer imprint is published by the registered company Springer Nature Switzerland AG.
The registered company address is: Gewerbestrasse 11, 6330 Cham, Switzerland

Preface

With the software-defined networking (SDN), network function virtualization (NFV), and fifth-generation wireless systems (5G) development, the global network is undergoing profound restructuring and transformation. However, with the improvement of the flexibility and scalability of the networks, the ever-increasing complexity of networks makes effective monitoring, overall control, and optimization extremely difficult. Recently, adding intelligence to the control plane through AI and ML becomes a trend and a direction of network development.

In this book, we apply different machine learning approaches to investigate solutions for intelligent monitoring, overall control, and optimization of networks. We focus on four scenarios of successfully applying machine learning in network space.

Intelligent Network Awareness In the network, different applications produce various traffic types with diverse features and service requirements. Therefore, in order to better manage and control networking, the intelligent awareness of network traffic plays a significant role. In Chap. 3, we discuss the main challenge of network traffic intelligent awareness and introduced several machine learning-based traffic awareness algorithms, such as traffic classification, anomaly traffic identification, and traffic prediction.

Intelligent Network Control Finding the near-optimal control strategy is the most critical and ubiquitous problem in a network. However, the traditional works on the control plane are largely relied on a manual process in configuring forwarding, which cannot be employed for nowadays network condition. To address this issue, several artificial intelligence approaches for self-learning control strategies in networks are introduced in Chap. 4.

Intelligent Network Resource Allocation Resource management problems are ubiquitous in the networking field, such as job scheduling, bitrate adaptation in video streaming, and virtual machine placement in cloud computing. Compared with the traditional with-box approach, ML methods are more suitable to solve the complexity of network resource allocation problems. Part of these works is introduced in Chap. 5.

Intent-Based Networking Management Future networks need to capture business intent and activate it network-wide in order to bridge the gap between the business needs and the network deliveries. Therefore, in this book, semantic comprehension function is introduced to the network to understand the high-level business intent. Part of these works is introduced in Chap. 6.

The topic of AI-driven networking is quite hot in current academia. We can now find many books on this topic. The following two features make this book a unique source for students and researchers. This book presents a four-tier network architecture named NetworkAI, which is applying AI and ML to deal with different level challenges of today's networking. This book presents a formalized analysis of several up-to-date networking challenges like virtual network embedding, QoS routing, etc. Some machine learning methods were introduced to effectively solve these challenges. These successful cases show how the ML can benefit and accelerate the network development.

Overall, this book aims at giving a comprehensive discussion on the motivation, problem formulation, and research methodology on applying machine learning in the future intelligent network. Although we made an earnest endeavor for this book, there may still be errors in the book. We would highly appreciate if you contact us when you find any.

Beijing, China — Haipeng Yao
Beijing, China — Chunxiao Jiang
Omaha, NE, USA — Yi Qian

Acknowledgments

Thanks to all the collaborators who have also contributed to this book: Tianle Mai, Mengnan Li, Chao Qiu, Peiying Zhang, Yaqing Jin, and Yiqi Xue.

Contents

Chapter 1
Introduction

1.1 Background

The current Internet architecture established from TCP/IP has gained huge success and been one of the indispensable infrastructures for our daily life, economic operation and society. However, burgeoning megatrends in the information and communication technology (ICT) domain are urging the Internet for pervasive accessibility, broadband connection and flexible management, which call for potential new Internet architectures. The original design tactic of the Internet, which is "Leaving the complexity to hosts while maintaining the simplicity of network", leads to the almost insurmountable challenge known as "Internet ossification": software in the application layer has developed rapidly, and abilities in the application layer have been drastically enriched. By contrast, protocols in the network layer lack scalability and the core architecture is hard to modify, which means that new functions have to be implemented through myopic and clumsy ad hoc patches in the existing architecture. For example, the transition from IPv4 to IPv6 is difficult to deploy in practice.

To improve the performance of the current Internet, novel network architectures have been proposed by the research communities to build the future Internet, such as Content Centric Networking (CCN) and Software-Defined Networking (SDN). Specifically, the past few years have witnessed a wide deployment of software defined networks. SDN is a paradigm, which separates the control plane from the forwarding plane, breaks vertical integration, and introduces the ability of programming the network. However, the work on the control plane largely relies on a manual process in configuring forwarding strategies. With the expansion of network size and the rapid growth of the number of network applications, current networks have become highly dynamic, complicated, fragmented, and customized. These requirements pose several challenges for tradition SDN.

Recently, AlphaGo's success comes at a time when researchers are exploring the potential of artificial intelligence to do everything from drive cars to financial

H. Yao et al., *Developing Networks using Artificial Intelligence*,
Wireless Networks, https://doi.org/10.1007/978-3-030-15028-0_1

market. Therefore, we ask if machine learning can provide a feasible alternative to a manual process for networking control and management. In other words, can software defined networking learn to manage resource on their own by machine learning technology? The recent success of using machine learning in many challenging domains suggest that this aim may not be impossible. Therefore, in this book, we focus on how machine learning can revolute networking development.

1.2 Overview of SDN and Machine Learning

In this section, we provide a more detailed view on software defined networking and machine learning.

1.2.1 Software Defined Networking (SDN)

SDN is a new type of network architecture. Its design concept is to separate the control plane of the network from the data forwarding plane, so as to realize the programmable control of the underlying hardware through the software platform in the centralized controller and realize flexible network resources. On-demand deployment. In the SDN network, the network device is only responsible for pure data forwarding, and can use common hardware; the operating system that is responsible for control will be refined into an independent network operating system, which is responsible for adapting different service characteristics, and the network operating system and Business characteristics and communication between hardware devices can be implemented programmatically.

As shown in Fig. 1.1, the SDN consists of three layers, forwarding plane, control plane and application plane.

Forwarding Plane Only responsible for network forwarding, not for the lowest infrastructure layer responsible for flow table-based data processing, forwarding, and state collection.

Control Plane The controller centrally manages all the devices on the network. The virtual entire network is a resource pool. The resources are flexible and dynamically allocated according to the different needs of users and the global network topology. The SDN controller has a global view of the network and is responsible for managing the entire network: for the lower layer, communicating with the underlying network through standard protocols; for the upper layer, providing the application layer with control over the network resources through the open interface.

Application Plane The top layer is the application layer, including various services and applications; the middle control layer is mainly responsible for processing the

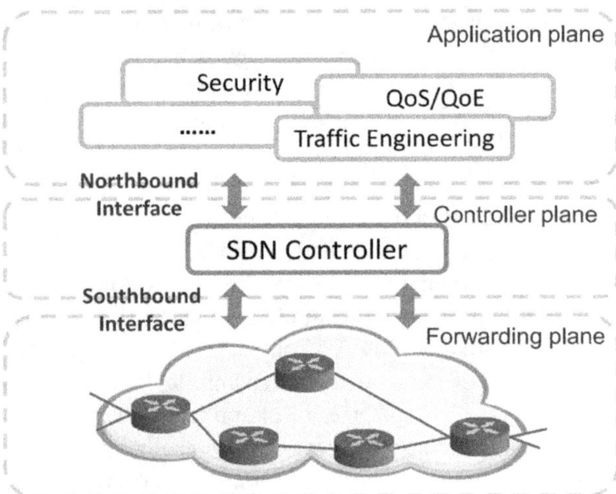

Fig. 1.1 The architecture of SDN

data plane resources, maintaining the network topology, state information, etc.; DN application layer through the programming layer provided by the programming interface The underlying device is programmed to open the control of the network to the user, and to develop various business applications based on the above, to achieve rich and colorful business innovation.

In addition, SDN includes two interfaces:

Southbound Interface The southbound interface is the channel through which the physical device and the controller transmit signals. The related device status, data flow entries, and control commands need to be communicated through the southbound interface of the SDN to implement device control.

Northbound Interface The northbound interface is an interface that is open to the upper layer service application through the controller. The purpose is to enable the business application to conveniently call the underlying network resources and capabilities. The service is directly applied to the service application, and the design needs to be closely related to the service application requirements.

The traditional network adopts a three-layer distributed network, a core layer, an aggregation layer and an access layer. Without a unified central control node, devices on the network learn the direct routing information advertised by other devices, so that each device decides how to forward it. This makes it impossible to control traffic from the perspective of the entire network. The typical architecture of SDN is divided into three layers: Forwarding plane, Control plane, and Application plane.

Therefore, compared with traditional networks SDN has the following advantages: first, the device hardware is normalized. The hardware only focuses on forwarding and storage capabilities and decouples from the service features. It can be implemented using a relatively inexpensive commercial architecture. Second, the intelligence of the network is all implemented by software. The types and functions of the network devices are determined by software configuration. The operation control and operation of the network is performed by the server as a network operating system (NOS). Third, the service response is relatively faster, and various network parameters such as routing, security, policy, QoS, traffic engineering, etc. can be customized, and configured in real time to the network, and the time for opening specific services will be shortened.

The essence of SDN is to enable users/applications to fully control the behavior of the network through software programming, to make the network software and agile. Its typical features are as followed. Control and forwarding are separated. The forwarding plane consists of controlled forwarding devices. The forwarding mode and service logic are controlled by the control application running on the separated control plane. An open interface between the control plane and the forwarding plane. SDN provides an open programmable interface to the control plane. In this way, the control application only needs to focus on its own logic, without having to pay attention to the underlying implementation details. Logically centralized control. The logically centralized control plane can control multiple forwarding plane devices, that is, control the entire physical network, thereby obtaining a global network state view, and implementing optimal network control according to the global network state view.

Based on the above points, SDN has the following advantages for the network as followed. First, SDN provides more flexibility for the use, control, and how to generate revenue for the network. Second, SDN has accelerated the introduction of new services. Network operators can deploy related functions through controlled software without having to wait for a device provider to add a solution to their proprietary device as before. Third, SDN reduces the operating expenses of the network and reduces the error rate. The reason is that the network is automatically deployed and the operation and maintenance fault diagnosis is realized, and the manual intervention of the network is reduced. Fourth, SDN helps to realize the virtualization of the network, thus realizing the integration of computing and storage resources of the network, and finally, the control and management of the entire network can be realized by a simple combination of software tools. Fifth, SDN makes the network and even all IT systems better oriented to business goals.

1.2.2 Machine Learning

Over the past few decades, machine learning techniques have attracted lots of attention from both academia and industry. Machine learning was proposed in the late 1950s as a key approach for Artificial Intelligence (AI) for the first time. The classical definition of ML can be described as "The development of computer models for training data that provides solutions to the problems of knowledge

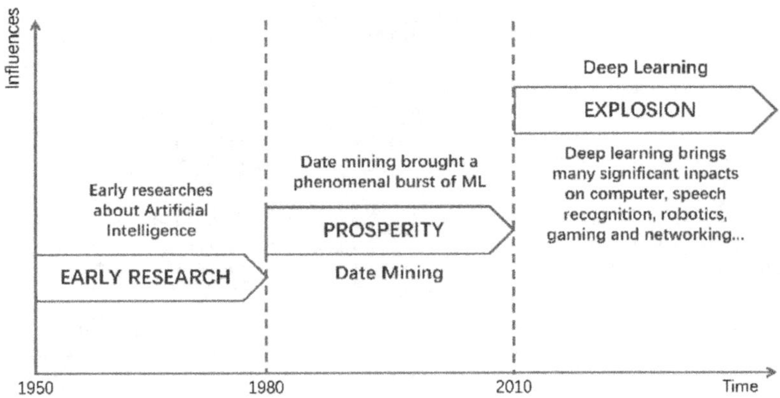

Fig. 1.2 Evolutions of machine learning techniques

acquisition and enhances the performance of developed systems without being explicitly programmed to".

We will have a look at machine learning's evolutions from the 1950s until now in Fig. 1.2. The first appearance of ML happened in the 1950s, and some early researches had been done from 1950 to 1980. With the emerging of data mining, there was a phenomenal burst of research and development in ML. Data mining is a group of algorithms about extracting the useful and unknown knowledge from datasets. Before this, ML was only used in classifying and predicting based on the known properties extracted from the training data, which needed the appropriate problem formulations, while data mining was used to discover the unknown properties from datasets. Data mining does not need to build an appropriate problem formulation but focuses on discovering valuable knowledge. During this period, mobile data communication technologies (e.g., LTE and LTE-A) developed rapidly, and the users' data exploded in both volume and speed, which enabled to be used to understand the performances and interests of users. Thus, the first practical applications appeared, such as advertisements and recommendations toward users' behaviors.

From the 2010s until now, an even more intelligent branch of ML, called as deep learning, has attracted the whole world's attention. If we need the novel insights from data to construct the computational models aiming at the high-level data abstractions, one tool for this is deep learning, hierarchical learning. Deep learning enables to use the unlabeled data to discover the intermediate or abstract representations in a hierarchical manner with deep neural networks. The higher-level features are defined based on the lower-level features to improve the classification or modeling results. Especially, there is a breakthrough when the distributed version of AlphaGo defeated the European Go champion Fan Hui for the first time. AlphaGo utilizes the Monte Carlo tree search algorithm and artificial neural networks to find the optimal moves.

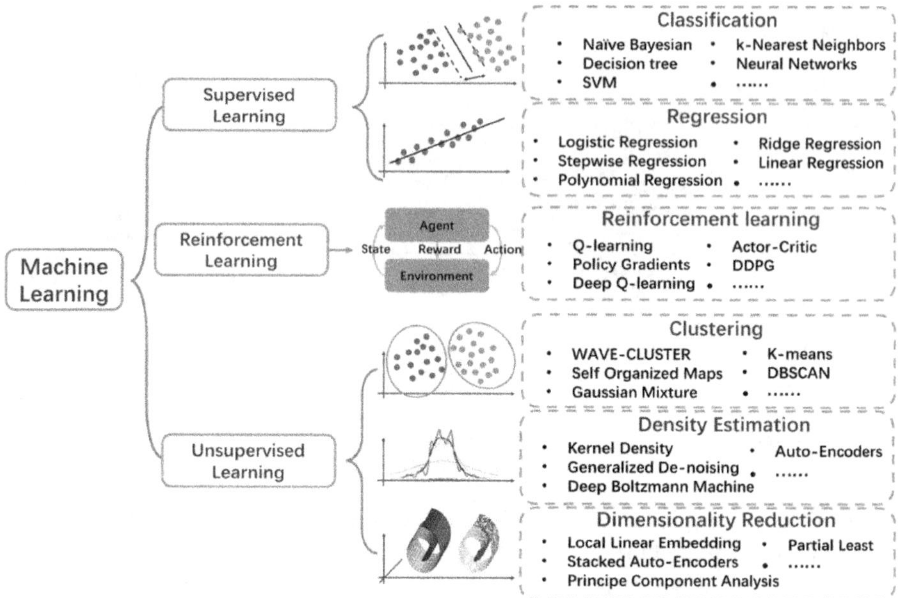

Fig. 1.3 Various machine learning algorithms and their corresponding classification and applications

After reviewing the evolution of machine learning, some items about ML algorithms will be presented. Existing machine learning techniques can be classified into supervised learning, unsupervised learning, and reinforcement learning. In supervised learning, algorithms need the labeled training datasets to construct the system models about the relations among inputs, outputs, and parameters. On the contrary, unsupervised learnings are not provided with labels and their main applications are to classify the sample sets into the different clusters based on the investigations of similarity between inputs. The third category, reinforcement learning, is another area of ML, where the agent takes actions in an environment to maximize the cumulative reward. Finally, in order to understand ML algorithms systematically, various ML algorithms, their corresponding classification and applications are illustrated in Fig. 1.3.

1.2.2.1 Supervised Learning

Supervised learning is the process of adjusting the parameters of a classifier to the desired performance using a set of samples of a known class. Supervised learning is a machine learning task that infers a function from the labeled training data. The training data includes a set of training examples. In supervised learning, each instance consists of an input object (usually a vector) and the desired output value (also known as a supervisory signal). The supervised learning algorithm analyzes the training data and produces an inferred function that can be used to map out new instances. An optimal solution would allow the algorithm to correctly determine the class labels of those invisible instances. This requires learning algorithms to be formed in a "reasonable" way from one training data to invisible.

The training set requirements for supervised learning include input and output, and can also be said to be features and goals. The goal of the training set is to be labeled (scalar) by the person. Under supervised learning, the input data is called "training data" and each set of training data has a clear identification or result. When establishing a predictive model, supervised learning establishes a learning process, compares the predicted results with the actual results of the "training data", and continuously adjusts the predictive model until the predicted outcome of the model reaches an expected accuracy.

Supervised learning is divided into classification algorithms and regression algorithms: For classification, the target variable is the category to which the sample belongs. In the sample data, the characteristics of each sample are included, for example, the training samples of various data input to the model are generated, and the data of one person is input to determine whether or not the patient has the data. As a result of cancer, the results must be discrete, with only "yes" or "no." For regression, regression is for predicting, for example, a training sample that inputs various data of a person into a model, and produces a result of "inputting one person's data and judging the future economic ability of the person 20 years later", and the result is continuous. Often get a regression curve. When the input arguments are different, the output dependent variables are not discretely distributed. The difference between classification and regression is that the target variable of the classification is discrete, and the target variable of the regression is continuous numerical.

1.2.2.2 Unsupervised Learning

Compared with supervised learning, the unsupervised learning training set has no artificially labeled results. In unsupervised learning, data is not specifically identified, and the learning model is used to infer some of the inherent structure of the data. Common application scenarios include learning of association rules and clustering. Common algorithms include the Apriori algorithm and the K-Means algorithm. The goal of this type of learning is not to maximize the utility function, but to find the approximate point in the training data. Clustering often finds fairly good visual classifications that match hypotheses. For example, aggregates based on demographics may form a rich aggregate in one group, as well as other poor aggregates.

Unsupervised learning seems very difficult: the goal is that we don't tell the computer what to do, but let it (computer) learn how to do something. There are two general ideas for unsupervised learning: The first idea is to not specify a clear classification when guiding the Agent but to adopt some form of incentive system when it succeeds. It should be noted that this kind of training is usually placed in the framework of decision-making problems because its goal is not to produce a classification system but to make the decision to maximize returns. This kind of thinking is a good summary of the real world, Agent can motivate those correct behaviors and penalize other behaviors.

Commonly used unsupervised learning algorithms mainly include principal components analysis methods such as PCA, equidistant mapping method, locally linear embedding method, Laplacian feature mapping method, black plug local linear embedding method and local tangent spatial permutation method. A typical example of unsupervised learning is clustering. The purpose of clustering is to bring together similar things, and we don't care what this class is. Therefore, a clustering algorithm usually only needs to know how to calculate the similarity and can start working.

1.2.2.3 Reinforcement Learning

Reinforcement learning is a reward-guided behavior in which the agent learns in a "trial and error" manner, and the goal is to maximize the reward of the agent through the interaction with the environment. Reinforcement learning is different from the supervised learning in connected learning. It is mainly reflected in the teacher's signal. The enhanced signal provided by the environment in reinforcement learning is to evaluate the quality of the action (usually a scalar signal), rather than telling the reinforcement learning system RLS (reinforcement learning system) how to generate the correct action. Because the external environment provides very little information, RLS must learn on its own experience. In this way, RLS gains knowledge in an action-evaluation environment and improves the action plan to adapt to the environment.

Reinforcement learning is developed from theories of animal learning, parameter disturbance adaptive control, etc. The basic principles are: If a certain behavioral strategy of the Agent leads to a positive reward (enhanced signal), then the trend of the Agent to generate this behavioral strategy will be strengthened. The goal of the Agent is to find the optimal strategy in each discrete state to maximize and reward the desired discount.

Reinforcement learning regards learning as a process of tentative evaluation. The Agent selects an action for the environment. After the environment accepts the action, the state changes, and at the same time, an enhanced signal (a reward or punishment) is sent to the Agent. The agent re-intensifies the signal and the current state of the environment. Choosing the next action, the principle of choice is to increase the probability of being positively strengthened (award). The selected action not only affects the immediate enhancement value but also affects the state of the environment at the next moment and the final enhancement value.

1.3 Related Research and Development

1.3.1 3GPP SA2

3GPP SA2 set up a research project "Study of enablers for Network Automation for 5G (eNA)" on 5G network intelligence at the Hangzhou conference in May 2017. The project belongs to 3GPP Rel 16, SA2 has officially discussed the project at the Gothenburg meeting in January 2018.

The project background of the project is that 3GPP SA2 introduces Network Data Analysis Function (NWDAF) in Rel 15's 5G core network. At present, the main application scenario of the function is a slice network, which provides network slice state analysis results to the policy control function and the network slice selection function by automatically analyzing the network data related to the network slice. On the other hand, in the 5G network architecture research of Rel 15, SA2 introduced some new requirements, such as on-demand mobility management, non-standardized QoS, traffic offloading and offloading. Without network data analysis, it is difficult to actually deploy and implement these requirements. Therefore, in order to make the 5G network more flexible and intelligent, Huawei has led the establishment of the eNA project in SA2.

The research goal of the eNA project is to collect and analyze network data through NWDA, generate analysis results, and then use the analysis results for network optimization, including customized mobility management, 5G QoS enhancement, dynamic traffic grooming and offload, and user plane functions. Select, based on the service usage of the UE, traffic policy routing and service classification.

1.3.2 ETSI ISG ENI

Recently, the European Telecommunications Standards Institute ETSI officially approved the establishment of a new industry standard group ENI (Experiential Networked Intelligence). The working group's goal is to align the scenarios and requirements of the network intelligence, define a closed-loop control model based on "observation-judgment-decision-action" and simulate human brain decision-making, and continuously optimize the network environment and continuously optimize decision-making results. Means to effectively cope with complex network management and control challenges and improve network operation management efficiency and experience.

The research scope of the working group includes analyzing the operational and operation and maintenance requirements of operators' traditional networks and SDN/NFV networks and guiding operators to build an architecture including adaptive sensing, flexible policy definition support, and intelligent decision-making and execution. The architecture should fully support operators' flexible business strategies and intelligent, self-optimizing, and autonomous smart network concepts.

At the same time, the architecture will also promote the development of technologies such as telemetry remote sensing, big data collection and management, and machine learning algorithms to support intelligent analysis and decision making. The unified strategy model is one of the core research technologies of network artificial intelligence.

The current ENI work process is divided into two phases: the first phase defines and describes the use cases and requirements, and makes them agree, analysing the gap between use cases and requirements and liaising with relevant standards groups. In the second phase, ENI will define the corresponding network architecture based on the results of the first phase, and design the use cases and requirements defined in the first phase. Currently, ENI has launched four related technical manuscript projects: use cases, requirements, context-aware strategy management gap analysis, terminology. At present, the drafting of these four manuscripts has come to an end, and the manuscript will be released after being reviewed. At the same time, ENI launched the architecture project in January 2018.

1.3.3 ITU-T FG-ML5G

During the ITU-T SG13 meeting in November 2017, representatives of business experts from different countries such as Germany, South Korea, China, Tunisia, and Africa proposed the establishment of a machine learning-network focus group, which was officially approved by the SG13 plenary discussion. The official name of the focus group is Machine Learning for Future Networks including 5G, referred to as FG-ML5G.

The FG-ML5G is a working platform open to both ITU and non-ITU members. The goal is to analyze how to apply machine learning to future networks, especially 5G networks, to improve network performance and user experience. Specific work includes analyzing and collaborating with current industry standards organizations; researching machine learning scenarios, potential needs, architectures, and specific interfaces, protocols, algorithms, data structures, and personal information protection in future networks; analyzing machine learning Impact on autonomous network control and management.

FG-ML5G is the focus group under SG13. It has three working groups internally to study application scenarios and business requirements, data format, machine learning technology, and network architecture. The research period is 1 year, and it can be determined by SG13 if necessary. Whether to postpone. The research report and draft standard of the FG-ML5G output can be used as input for subsequent research on SG13.

1.4 Organizations of This Book

This book is organized as depicted in Fig. 1.4. We first propose the concept of NetworkAI, a novel paradigm that applying machine learning to automatically control a network. NetworkAI employs reinforcement learning and incorporates network monitoring technologies such as the in-band network telemetry to dynamically generate control policies and produces a near optimal decision. We employ the SDN and INT to implement a network state upload link and a decision download link to accomplish a closed-loop control of a network and build a centralized intelligent agent aiming at learning the policy by interaction with a whole network.

Then, we discuss the possible machine learning methods for network awareness. With the rapid development of compelling application scenarios of the networks, such as 4K/8K, IoT, it becomes substantially important to strengthen the management of data traffic in networks. As a critical part of massive data analysis, traffic awareness plays an important role in ensuring network security and defending traffic attacks. Moreover, the classification of different traffic can help to improve their work efficiency and quality of service (QoS).

Furthermore, we discuss how machine learning can achieve network automatically control. Finding the near-optimal control strategy is the most critical and ubiquitous problem in a network. Examples include routing decision, load balancing, QoS-enable load scheduling, and so on. However, the majority solutions of these problems are largely relying on a manual process. Therefore, to address this

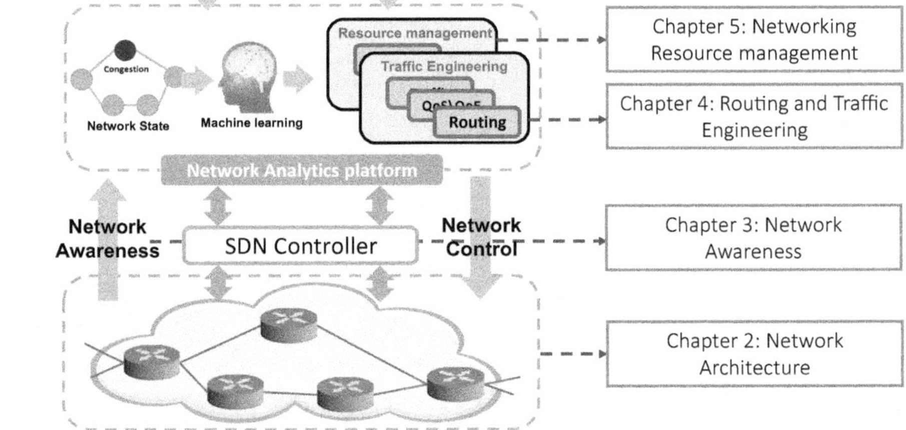

Fig. 1.4 Book organization

issue, in Chap. 4, we apply several artificial intelligence approaches for self-learning control strategies in networks.

In addition, resource management problems are ubiquitous in the networking field, such as job scheduling, bitrate adaptation in video streaming and virtual machine placement in cloud computing. In Chap. 5, we focus on how machine learning can optimal resource management decision.

Finally, in order to reduce network complexity and improves automation by eliminating manual configurations, we discuss the intent-based networking which allows a user or administrator to send a simple request—using natural language—to plan, design and implement/operate the physical network. For example, an IT administrator can request improved voice quality for its voice-over-IP application, and the network can respond. For intent-based networking, the translation and validation system take a higher-level business policy (what) as input from end users and converts it to the necessary network configuration (how) by natural language understanding technology. To achieve intent-based networking, in Chap. 6, we focus on how machine learning can technology can be used in the natural language understanding in translation and validation system.

1.5 Summary

In this chapter, we mainly introduce the background. We first introduce the motivation of this book. Based on the motivation, we propose the coordinated network architecture and discuss the key technologies and challenges in the architecture. Then, the related research and development are provided. Finally, we give the organizations of this book.

Chapter 2
Intelligence-Driven Networking Architecture

Network architecture provides a full picture of the established network with a detailed view of all the resources accessible. The architecture of software defined network facilitates separation of the control plane from the forwarding plane. However, the work on the control plane largely relies on a manual process in configuring forwarding strategies. To address this issue, in this chapter, we propose NetworkAI, a new network architecture exploiting software-defined networking, network monitor technologies and reinforcement learning technologies for controlling networks in an intelligent way. NetworkAI implements a network state upload link and a decision download link to accomplish a closed-loop control of the network and builds a centralized intelligent agent aiming at learning the policy by interacting with the whole network.

2.1 Network AI: An Intelligent Network Architecture for Self-Learning Control Strategies in Software Defined Networks

Recently, Software Defined Networking (SDN) has received a large amount of attention from both academia and industry. The SDN paradigm decouples control plane from data plane and provides a logically-centralized control plane, wherein the network intelligence and state are logically centralized, and the underlying network infrastructure is abstracted for the network applications and services [1, 7]. This logically-centralized control mechanism provides the efficient use of network resources for network operators and the programmability brought by SDN simplifies the configuration and management of networks. Therefore, network operators can easily and quickly configure, manage and optimize network resource in SDN architecture [3]. However, even though a separation of the control plane from the forwarding plane facilitates large scale and dynamic networks, a challenge that

H. Yao et al., *Developing Networks using Artificial Intelligence*,
Wireless Networks, https://doi.org/10.1007/978-3-030-15028-0_2

remains un-tackled is that the work on the control plane relies heavily on a manual process in configuring forwarding strategies.

Finding the near-optimal control strategy is the most critical and ubiquitous problem in a network. The majority of approaches to solve this problem today usually adopted the white-box approaches [5, 6]. With the expansion of network size and the rapid growth of the number of network applications, current networks have become highly dynamic, complicated, fragmented, and customized. These requirements pose several challenges when applying these traditional white-box algorithms [4]. Specifically, a white-box approach generally requires an idealized abstraction and simplification of the underlying network; however, this idealized model often poses difficulties when dealing with a real complex network environment. In addition, the white-box method presents poor scalability under different scenarios and applications.

Owing to the success of Machine Learning (ML) related applications, such as robotic control, autonomous vehicles, and Go [8], a new approach for network control through ML has emerged. This new networking paradigm using is firstly proposed by Mestres et al., which is referred to as Knowledge-Defined Networking (KDN) [9]. However, KDN and some other similar works [10, 12] just proposed some concept, no detail was described in these papers and no actual work has been implemented.

In this part, we propose NetworkAI, an architecture exploiting software-defined networking, network monitor technologies (e.g., traffic identification, In-band Network Telemetry(INT)), and reinforcement learning technologies for controlling networks in an intelligent way. NetworkAI implements a network state upload link and a decision download link to accomplish a close-loop control of network and builds a centralized intelligent agent aiming at learning the policy by interacting with the whole network. The SDN paradigm decouples control plane from data plane and provides a logically-centralized control to whole underlying network. Some new network monitor technologies, such as In-band Network Telemetry(INT), [2, 11, 13] can achieve millisecond uploading of the network state and provide real-time packet and flow-granularity information to a centralized platform [14]. In addition, a network analytical platform, such as the PNDA [15], provides big data processing services via some technologies such as Spark and Hadoop. SDN and monitor technologies offer a completely centralized view and control to build the interaction framework of a network, thus enabling the application of ML running in a network environment to address the network control issues.

NetworkAI applies DRL to effectively solve real-time large-scale network control problems without relying on too much manual process and any assumptions of the underlying network. RL involves agents that learn to make better decisions from experiences by interacting with the environment [13, 17]. During training, the intelligent agent begins with no prior knowledge of the network task at hand and learns by reinforcement based on its ability to perform a task. Particularly, with the development of deep learning (DL) techniques, the success of combined application of RL and DL to large-scale system control problems (such as GO [8] and playing games) proves that the deep reinforcement learning (DRL) algorithm can deal with the complicated system control problem. DRL represents its control policy as a neural network that can transfer raw observations (e.g., delay, throughput, jitter) to the decision [16]. Deep learning (DL) can effectively compress the network state space thus enabling RL to solve large-scale network decision-making problems that were previously found difficult in handling high-latitude states and motion space.

In NetworkAI, the SDN and new network monitor technologies are employed herein to construct a completely centralized view and control for geographical distributed network and build a centralized intelligent agent to generate a network control policy via DRL. The NetworkAI can intelligently control and optimize a network to meet the differentiated network requirements in a large-scale dynamic network.

Different from the traditional white-box approaches, this part proposes a new network paradigm (i.e. NetworkAI) in applying ML to solve network control problem. The main contribution of this section can be summarized as follows:

- We employ the SDN and INT to implement a network state upload link and a decision download link to accomplish a close-loop control of a network and build a centralized intelligent agent aiming at learning the policy by interaction with a whole network.
- We apply DRL to effectively solve real-time large scale network control problems without too much manual process and any assumptions of underlying network, where the DRL agent can produce a near-optimal decision in real time.

2.1.1 Network Architecture

In this subsection, we firstly introduce the NetworkAI architecture and elaborate how it operates. The model of the NetworkAI is shown in Fig. 2.1. This model consists of three planes called the forwarding plane, the control plane, and the AI plane. And then, we will give a detailed description of each layer.

2.1.1.1 Forwarding Plane

The function of the forwarding plane is forwarding, processing, and monitoring data packets [18]. The network hardware, which is composed of line-rate programmable forwarding hardware, only focuses on simply data forwarding without embedding any control strategies. The control rules are issued by the SDN controller via southbound protocols such as OpenFlow [23] or P4 [24]. When a packet comes into the node, it will be forwarded and processed according to these rules. Besides, there are some monitoring processes embedded in nodes. The network monitor data will be collected and sent to the analytical platform. Thus, it can offer complete network state information to facilitate the AI plane to make decisions.

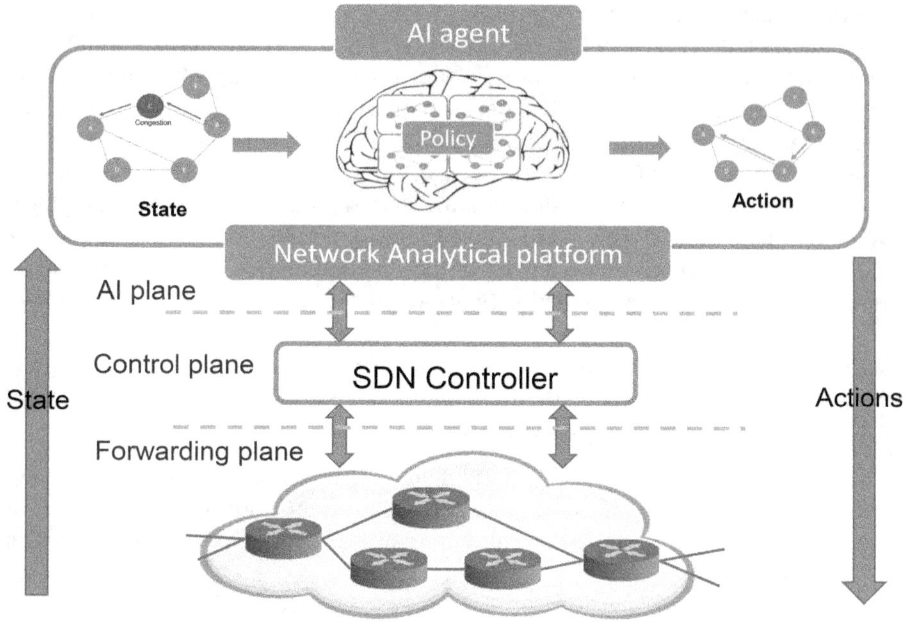

Fig. 2.1 The NetworkAI architecture

2.1.1.2 Control Plane

The function of the control plane is to connect the AI plane and the forwarding plane. This plane provides abstractions for accessing lower-level geographical distributed forwarding plane and pools the underlying resources (such as link bandwidth, network adapter, and CPU capacity) to the AI plane. The function of the SDN controller is to manage the network through standard southbound protocols and interact with the AI plane through the northbound interfaces. This logically-centralized plane eases the burden of the network control problem imposed by a geographical distributed network. Thus, the policies generated by the AI plane can be quickly deployed into the network.

2.1.1.3 AI Plane

The function of the AI plane is to generate policies. In the NetworkAI paradigm, the AI plane takes advantage of SDN and monitor techniques to obtain a global view and control of the entire network. The AI agent learns the policy through interaction with the network environment. While learning the policy is a slow process, the network analytical platform provides a big data storage and computing capacity. Fundamentally, the AI agent processes the network state collected by the forwarding plane, then transforming the data to a policy through RL and using that policy to make decisions and optimization.

2.1.2 Network Control Loop

In the NetworkAI architecture, we design the network state upload link and the decision download link to accomplish a close-loop control of network. The NetworkAI architecture operates with a control loop to provide an interactive framework for a centralized agent to automatically generate strategies. Now, in this subsection, we will detail how the NetworkAI architecture implement a control loop of the whole network and how the intelligent agent learns the policy by RL approach.

2.1.2.1 Action Issue

In traditional distributed networks, the control plane of the network node is closely coupled with the forwarding plane which has only partial control and view over the complete network. This partial view and control can lead to no global convergence of learning result. The AI agent need to continually converge to a new result when network state changed which will lead to a bad performance in real-time control problem. For purpose of achieving global optimum, the controlling and managing of the whole network is premise. SDN is a paradigm which separates the control plane from the forwarding plane and therefore breaks the vertical integration. The SDN controller treats the entire network as a whole. In this manner, the SDN acts as a logical-centralized agent to control the whole network. The SDN controller issues a control action through an open and standard interface (e.g., OpenFlow, P4). These open interfaces enable the controller to dynamically control heterogeneous forwarding devices, which is difficult to achieve in traditional distributed networks.

As demonstrated in Fig. 2.2, the agent can issue control actions to the forwarding plane via southbound protocols according to the decisions made at the AI plane, the network node at the forwarding plane operates based on the updated rules imposed by the SDN controller [9]. In this manner, we realized the global controllability of the entire network.

2.1.2.2 Network State Upload

In the SDN network architecture, the controller can send an action decision to the underlying network with an aim to acquire a complete network control. Furthermore, obtaining the complete real-time view of the whole network is also crucial to make near-optimal decisions. The most relevant data that should be collected is network state information and traffic information. To this end, we

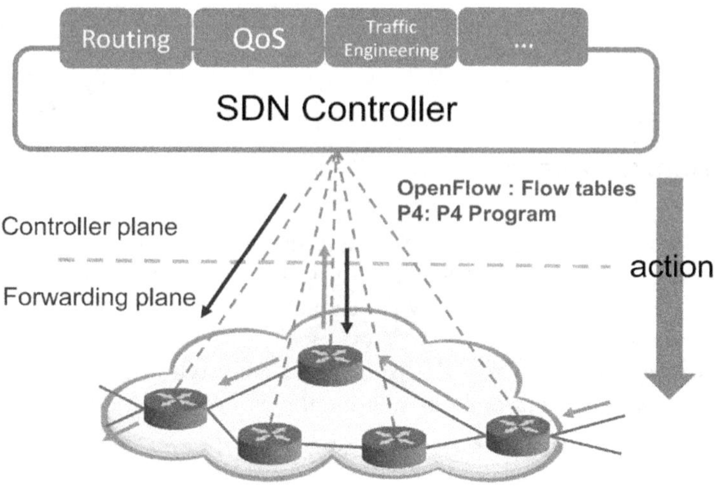

Fig. 2.2 The action issuing process

designed the upload link to access fine-grained network and traffic information. In this subsection, we will introduce how the NetworkAI architecture achieves network state upload.

1. **Network Information**: Network information mainly refers to the status of the network device (information below the L2 layer), including the network physical topology, hop latency, and queue occupancy. In our architecture, we borrow in-band telemetry technology to achieve fine-grain network monitoring.

   Obtaining fine grained network monitoring data of dynamic networks is a concern of NetworkAI. The traditional monitor technologies are commonly based on out-band approaches. In this way, monitoring traffic is sent as dedicated traffic, independent from the data traffic ("probe traffic"), such as SNMP, synthetic probes. These methods bring to much probe traffic in networking and overhead computation to control plane in large-scale dynamic networking which extremely degrades the performance of real-time controlling.

   In-band network telemetry is a framework designed to allow the collection and reporting of network state by the data plane, without requiring intervention or computation by the control plane. The core idea of INT is to write the network status into the header of a data packet to guarantee the granularity of monitoring the network at the packet level [25]. The telemetry data is straightforward adding to the packet. Therefore, the end-to-end monitoring data can be retrieved from the forwarding node through Kafka or IPFIX directly to AI plane's big data platform without intervening by the control plane.

   The INT monitoring technology model is illustrated in Fig. 2.3. From the Fig. 2.3, we can see that a source node embeds instructions in the packets, listing the types of network information needs to be collected from the network elements (e.g., hop latency, egress port TX link utilization, and queue occupancy). Each

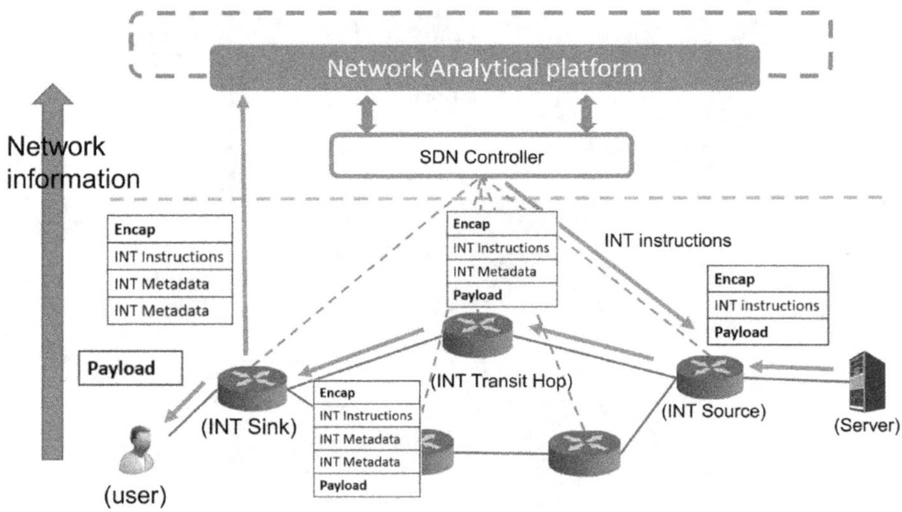

Fig. 2.3 The collection process of network information

network element inserts the requested network state in the packet as it traverses the network. When the packet is sent to the INT Sink, the load data is sent to the user and the telemetry data is sent to the network analytical plane.

Data collection can be realized based on the actual traffic. INT provides the ability to observe and collect real-time and end-to-end network information across physical networks. In addition, the mechanism of INT vanishes the overhead communication of probe traffic and overhead computation of control plane. Due to borrowing from INT technology, the AI plane can obtain millisecond fine grain network telemetry data, which gives the possibility to react network in time.

2. **Traffic Information**: Traffic information mainly include service-level information (e.g., QoS/QoE), anomaly traffic detection information(e.g., elephant flow), etc. In network, different applications produce various traffic types with diverse features and service requirements. In order to better manage and control networking, the identification of network traffic plays a significant role [26]. For instance, elephant flow is an extremely large continuous flow established by a TCP (or other protocol) flow [27]. Elephant flows can occupy network bandwidth and bring seriously congestion to the network. Therefore, it is of great significance for AI plane to detect or even predict the elephant flow in time and take the necessary action to avoid network congestion.

   In our architecture, as illustrated in Fig. 2.4, several monitor processes embedded in some nodes to transfer the raw traffic data (e.g., flow granularity data, relevant traffic feature, and Deep Packet Inspection (DPI) information) to the traffic information via data mining methods, such as traffic classification and traffic anomaly detection [28, 30]. Then, the traffic information will be upload to the network analytical plane to assist AI plane in decision making.

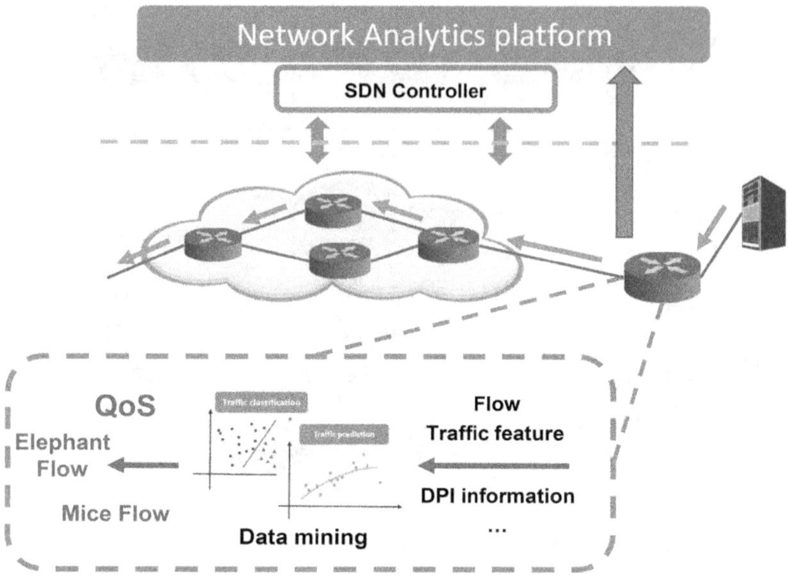

Fig. 2.4 The collection process of traffic information

2.1.2.3 Policy Generation

With the aim to apply ML method to realize an intelligent network control, we have already constructed the interaction framework between an AI agent and the network environment. In this part, we will describe how to use ML method to generate the network policy. The machine learning methods contain three approaches, supervised learning, unsupervised learning and reinforcement learning. Compared to supervised and unsupervised learning, reinforcement learning is more suitable for close-loop control problems. In particular, with the development of DL, the success of combining DL and RL for applications in decision-making domains (Playing Atari with Deep Reinforcement Learning by DeepMind at NIPS 2013 and the 2016 Google AlphaGo success on Go) demonstrates that DRL can effectively solve large-scale system control problems. Thus, we apply RL method to solve the large-scale network control problem.

RL based learning tasks are usually described as a Markov decision process, as shown in Fig. 2.5. At each step, an agent observes the current state s_t from the network environment and the agent takes an action at according to a policy $\pi(a|s)$. Following the action, the network environment transfer to state $s_t + 1$ and the agent observes a reward signal r_t from environment. The goal of the reinforcement learning is to obtain the optimal behavior policy that maximizes the expected long-term reward. Specifically, in the network scenario, the state is represented by the network state and flow information, the action is by network behaviors (e.g., CDN selection, routing selection) and the reward is based on the optimal target.

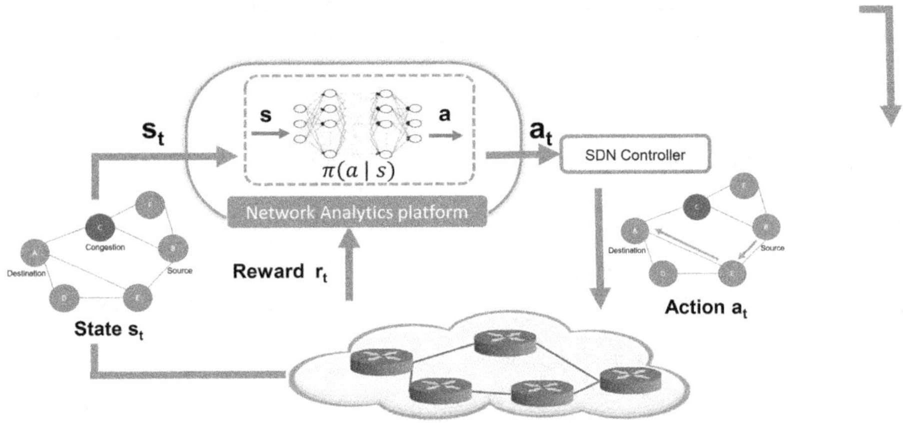

Fig. 2.5 The reinforcement learning process

The RL agent uses a state-action value function $Q(s, a)$ to measure the actions expected for a long-term reward on the state s. Starting from a random Q-function, in q-learning algorithm, the agent continuously updates its Q-values by:

$$Q(s_t, a_t) \overset{\alpha}{\leftarrow} r_{t+1} + \lambda Q(s_{t+1}, a_{t+1}) \tag{2.1}$$

where $x \overset{\alpha}{\leftarrow} y \equiv x \leftarrow x + \alpha(y - x)$ and λ is the discount parameter. Using these evolving Q-values, the agent chooses the action with the highest $Q(s, a)$ to maximize its expected future rewards.

Particularly, the combination of DL and RL takes a step further in complex system control. The traditional RL algorithm records the reward of $(state, action)$ through the table-based method, which will lead to complexity issues that the RL is not designed for, namely memory complexity, computational complexity, and sample complexity, as its use was inherently limited to low-dimensional problems [31]. Specifically, in a large-scale and highly dynamic network, there are too many $(state, action)$ pairs. It is often impractical to maintain the Q-value for all possible $(state, action)$ pairs. Hence, it is common to use a parameterized function $Q(s, a; \theta)$ to approximate $Q(s, a)$. Deep neural networks have powerful function approximation and representation learning capabilities [32]. The DL algorithm can automatically extract low-dimensional features from high-dimensional data. Therefore, DL can effectively compress the network state space, as illustrated in Fig. 2.6, thus enabling RL to solve large-scale network decision-making problems that were previously found difficult in handling high-latitude states and motion space.

Based on such reinforcement learning framework, the data flow in NetworkAI is described as follow. The network monitor data and traffic information will be collected by the upload link, the decision for each flow calculated in AI plane send to SDN controller via northbound Interface. The SDN control then issue the

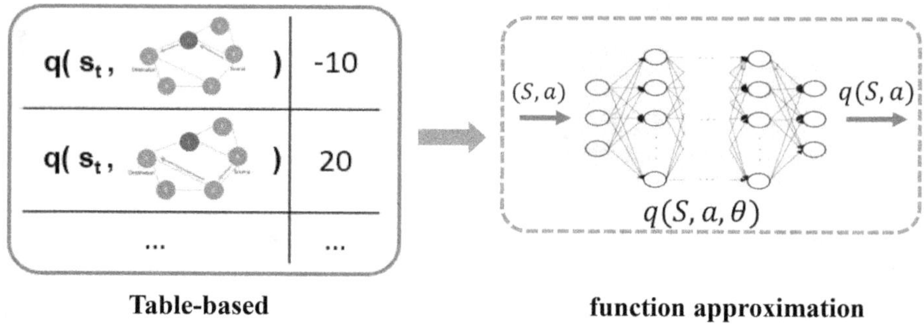

Fig. 2.6 The compress process of mass states

rule through southbound interface. Thus, this data flow achieves RL agent through interaction with the underlying network. Different applications just need craft the reward signal to guide the agent toward good policy to meet their objectives.

In our architecture, we apply deep reinforcement learning to generate network policy. Combining RL with DL leads to a general artificial intelligence solution for solving complex network control problems. We believe that introducing DRL for network decision making presents two main advantages.

First, the DRL algorithm is a black-box approach. The DRL agent only need to have different network decision tasks and optimization goals in designing action spaces and rewards without changing the mathematical model. In addition, because an artificial neural network has the characteristic of expressing arbitrary nonlinear mappings, the DRL agent can understand a nonlinear, complex, multi-dimensional network control problem without the need of any simplifications. On the contrary, traditional white-box approaches require assumptions and simplifications of the underlying network aiming at building the equivalent mathematical model and tailoring for a problem that has to be optimized.

Second, the DRL agent does not need to converge again when network state changed [22]. Once the DRL agent trained, an approximate optimal solution can be calculated in single step through matrix multiplication where the time complexity is only approximately $O(n^2)$, where n is represented by number of network nodes. In contrast, the heuristic algorithms need take many steps to converge to a new result, where leads to high computational time cost. For example, the time complexity of ant colony algorithm is $O(n \times (n-1) \times mt)$, where n is represented by number of network nodes , m is number of ants , t is number of iterations. Therefore, DRL offers a tremendous advantage for the real-time control of a dynamic network.

Above all, the NetworkAI achieves applying RL approach for the real-time control of the network. The SDN, INT and traffic identification technologies are used to implement the network state upload link and the decision download link respectively with an aim to obtain a centralized view and control of a complex network systems. In addition, DRL agent in AI plane can effectively solve complexity network control problem without the need of any simplifications of real network environment. Furthermore, that a near-optimal solution can quickly be calculated once it is trained represents an important advantage for the real-time control of the network. Thus, the NetworkAI facilitates an automated control of a large-scale network.

2.1.3 *Use Case*

The main objective of this use case is to demonstrate that it is possible to model the behavior of a network with the proposed NetworkAI architecture. In particular, we present a simple example in the context of QoS routing, where the NetworkAI was used to make intelligent decisions to select the best routing path aiming at satisfying the QoS requirements.

The traditional Internet design is based on end-to-end arguments with an aim to minimize the network supports. This type of architecture is perfectly suited for data transmission where the primary requirement is reliability [10]. However, with the proliferation of various applications (such as multimedia transmission application, where timely delivery is preferred over reliability), the demands of each application are different. Thus, the network should support QoS in a multi-application traffic scenario [21]. However, the question of how to support end-to-end QoS is an on-going problem.

QoS routing mainly involves path selection that meets the QoS requirements arising from different service flows. It is a mechanism of routing based on QoS requests of a data flow and the available network resources. The typical QoS indicators for applications in the network are different, as demonstrated in Table 2.1 in which we list the QoS requirements for several applications.

The dynamic QoS routing can be seen as a Constrained Shortest Path (CSP) problem, which is an NP-complete problem [33]. Although researchers from both academia and industry have proposed many solutions to solve the QoS limitations of the current networking technologies [19, 20, 33], many of these solutions either failed or were not implemented because these approaches come with many challenges. The tradition heuristic methods bring high computational time cost

Table 2.1 The QoS indicator for several applications

	QoS			
Application	Delay	Throughput	Jitter	Losses
Realtime multimedia	√	√	√	
Augmented reality or virtual reality	√	√	√	√
VoIP	√		√	
Scheduling in datacenters	√	√	√	√
Internet of vehicles	√			√

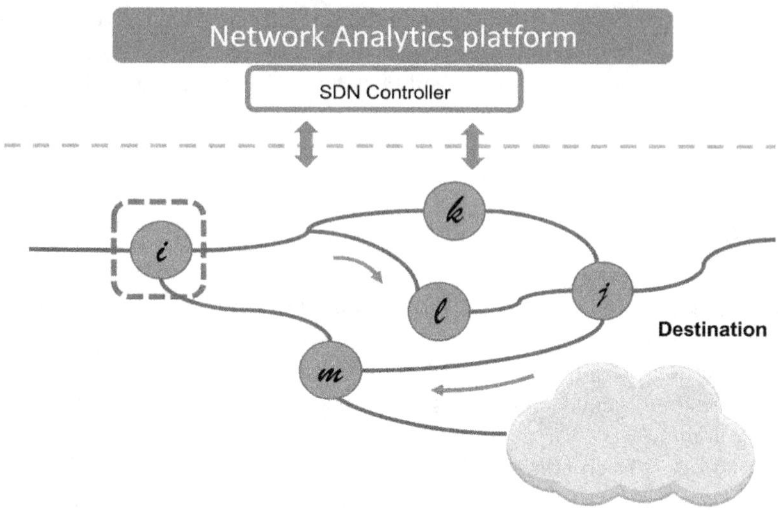

Fig. 2.7 The experiment example

in network control, which is difficult to apply to real-time dynamic network environment.

The NetworkAI paradigm can address many of the challenges posed by the QoS-Routing problem. RL approach can calculate an near-heuristic solution in one step, which can benefit for real-time control to large-scale network. And the closed-loop architecture in NetworkAI provide an interaction framework which achieves applying DRL in geo-distributed network. In our experiment, we applied the Deep Q-learning algorithm in a central agent, aimed to select the best routing path to minimize the network delay [21, 29, 34].

Methodology We simulated a network with 36 nodes and 50 full-duplex links, with uniform link capacities and different transmission delay. One of these nodes applied DQN and the others run short path algorithm. In addition, we devised ten traffic intensity (TI) levels, ranging from 1 to 10, which represent the volumes of the network traffic at different times. The traffic generated randomly in each node and volumes is subject to poisson distribution.

As shown in Fig. 2.7, it is a part of the whole experiment topology. We applied DQN to guide node i's routing action. In our experiment, state is represented by the link delay, node processing delay, which is up to 86^{20} states. Action is represented by the tuple of node which univocally determine the paths for the source-destination node pairs. As shown in Fig. 2.7, the action space for the node i to destination j is (i, k, j), (i, l, j) and (i, m, j). In addition, reward is represent by the delay from the source to the destination node.

In our experiment, we employed a neural network that has five fully connected hidden layers with a sigmoid activation function as well as a trained DQN on the gym platform [35] and Keras [36]. In addition, we devised ten traffic intensity (TI)

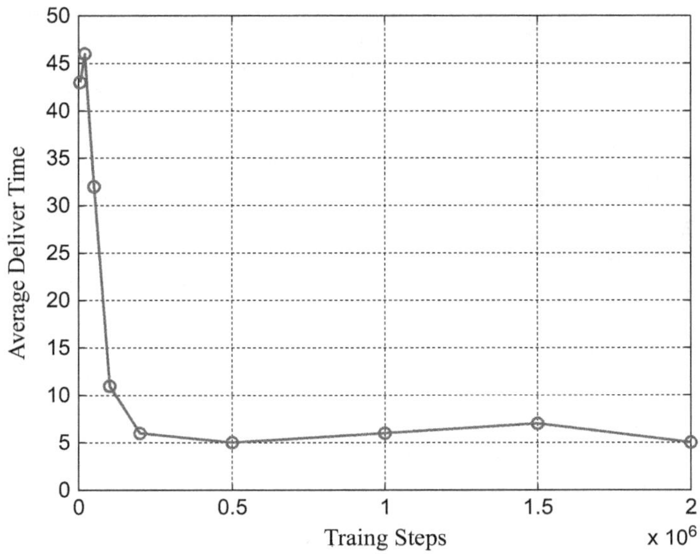

Fig. 2.8 The DRL learning process

levels, ranging from 1 to 10, which represent the volumes of the network traffic at different times. The DRL agent was trained for 200K steps for each TI.

Experimental Results and Analysis The DRL learning process is demonstrated in Fig. 2.8. This relevant outcome is that the DRL performance increases with training steps and the DRL agent be convergence when the training step more than 200K. The reason is that the DRL learning is a process that the result is close to a near-optimal strategy by interacting with the environment. The second simulation results are demonstrated in Fig. 2.9. It can be seen that the average transmission time of the network increases with the increase of network traffic load. When the load level is low, with the increase of network load, the increase of average transmission time is stable. But when the network load continues to increase, the average transmission time increases sharply, due to the fact that the network capacity is close to saturation.

In our experiment, the benchmark algorithm is the shortest path algorithm. When the network load is low, there is no congestion in the network and the shortest path is the optimal path. So, the benchmark result is better. With the increase of network load, congestion occurs on the shortest path of the network, and the network agent will have to choose non-congested links for transmission. Thus, in this situation, the DRL performance performs much better than the benchmark.

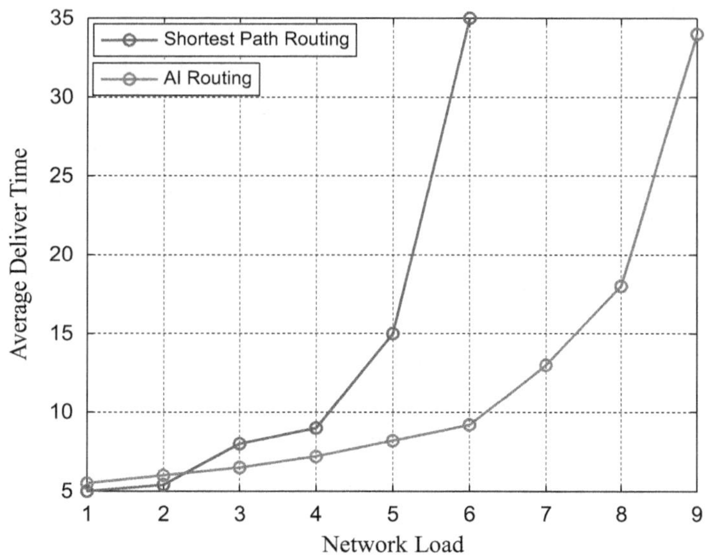

Fig. 2.9 The average delivery time over different network loads

2.1.4 Challenges and Discussions

In this subsection, we proposed a new network paradigm named NetworkAI. We introduce this architecture to effectively solve the two main problems when apply ML in network, how to run ML on distributed network and how to generate complex strategies. Through the technologies of SDN, monitor technologies, and RL, the intelligent agent can automatically generate network strategies for network control and management. Moreover, through a use case, we have demonstrated that a fully automated DRL agent can provide routing policy that tends to minimize the network delay.

Compared with the traditional network architectures, we apply SDN and monitor technologies to implement a completely centralized view and control for distributed network systems and build a closed control loop for real-time upload of network state info and download of actions.

At the same time, DRL is introduced to effectively generate control policy in network systems. Benefit from the DL method with powerful function approximation and representation learning capabilities, the massive state space of the network can be effectively compressed. In addition, due to the black-box feature of RL approach, for different network decision tasks and optimization goals, only the action space and reward have to be designed. Furthermore, the near-optimal solution can be calculated in single step once the DRL trained. Compared with the traditional heuristic algorithms, this feature presents a huge advantage for real-time network applications.

The NetworkAI paradigm brings significant advantages to networking. However, it also meets some challenges that need to be addressed further.

2.1.4.1 Communication Overhead

The communication overhead for retrieving and issuing data is a serious problem in SDN architecture. While the convenience brought by centralized framework, it leads to too much interaction between centralized controller and distributed forwarding unit. The performance of NetworkAI will be degraded as a result of the rapid flow table update to all forwarding unit. To address it, NetworkAI can borrow some technologies from SDN. One possible solution is segment routing technology, which implements the source routing and tunnel method to effectively reduce the flow table update. Another way to alleviate this problem is to employ a cluster of controller to handle larger flow tables [37].

2.1.4.2 Training Cost

RL approach provides flexible tools to address network control problems. Nonetheless, RL method involve a large amount of training cost, especially in network scenario where exists mass applications and services. When new businesses or services appears, the agent requires a significant level of training cost which weaken the flexibility of NetworkAI. In this context, the NetworkAI paradigm requires mechanisms to reduce the training cost to satisfy mass of new network applications. A notable approach to address this problem is adding prior knowledge to accelerate the training process, such as transfer learning, imitation learning [38]. In this sense, these trick may reduce the training cost when new businesses or services appear and essentially in taking a step further in performance of NetworkAI.

2.1.4.3 Testbeds

To evaluate the performance of new network designs and algorithms, testbeds are more convincing for network experiments compared with simulators and emulation platforms, because testbeds can incorporate real traffic and real network facilities [39]. Building a complex experimental environment will be the most critical issue for applying AI in a network. In particular, due to the fact that the NetworkAI architecture is aimed at a complex, highly dynamic multi-application network environment, it is difficult to obtain convincing experiments through the network simulator. Therefore, in the immediate future, we plan to build a large-scale real NetworkAI testbed to expand our experiments.

2.2 Summary

In this chapter, we presents NetworkAI, an intelligent architecture for self-learning control strategies in SDN networks. NetworkAI employs deep reinforcement learning and incorporates network monitoring technologies such as the in-band network telemetry to dynamically generate control policies and produces a near optimal decision.

References

1. O. N. Foundation, "Software-defined networking: The new norm for networks," 2012.
2. C. Qiu, F. R. Yu, H. Yao, C. Jiang, F. Xu, and C. Zhao, "Blockchain-based software-defined industrial internet of things: A dueling deep q-learning approach," *IEEE Internet of Things Journal*.
3. S. Sezer, S. Scott-Hayward, P. K. Chouhan, B. Fraser, D. Lake, J. Finnegan, N. Viljoen, M. Miller, and N. Rao, "Are we ready for sdn? implementation challenges for software-defined networks," *IEEE Communications Magazine*, vol. 51, no. 7, pp. 36–43, 2013.
4. X. Lei, C. Jiang, N. He, H. Zhu, and A. Benslimane, "Trust-based collaborative privacy management in online social networks," *IEEE Transactions on Information Forensics & Security*, vol. 14, no. 1, pp. 48–60, 2018.
5. H. E. Egilmez, S. T. Dane, K. T. Bagci, and A. M. Tekalp, "Openqos: An openflow controller design for multimedia delivery with end-to-end quality of service over software-defined networks," in *Signal & Information Processing Association Summit and Conference*, pp. 1–8, 2012.
6. S. Yussof and O. H. See, "The effect of ga parameters on the performance of ga-based qos routing algorithm," in *International Symposium on Information Technology*, pp. 1–7, 2008.
7. H. Yao, B. Zhang, P. Zhang, S. Wu, C. Jiang, and S. Guo, "Rdam: A reinforcement learning based dynamic attribute matrix representation for virtual network embedding," *IEEE Transactions on Emerging Topics in Computing*.
8. D. Silver, A. Huang, C. J. Maddison, A. Guez, L. Sifre, G. V. D. Driessche, J. Schrittwieser, I. Antonoglou, V. Panneershelvam, and M. Lanctot, "Mastering the game of go with deep neural networks and tree search," *Nature*, vol. 529, no. 7587, p. 484, 2016.
9. A. Mestres, A. Rodrigueznatal, J. Carner, P. Barletros, E. Alarc0́n, M. Solé, V. Muntés, D. Meyer, S. Barkai, and M. J. Hibbett, "Knowledge-defined networking," *Acm Sigcomm Computer Communication Review*, vol. 47, no. 3, pp. 2–10, 2016.
10. H. Yao, C. Qiu, C. Fang, X. Chen, and F. R. Yu, "A novel framework of data-driven networking," *IEEE Access*, vol. 4, no. 99, pp. 9066–9072, 2017.
11. X. Lei, C. Jiang, N. He, Q. Yi, and J. Li, "Check in or not? a stochastic game for privacy preserving in point-of-interest recommendation system," *IEEE Internet of Things Journal*, vol. PP, no. 99, pp. 1–1, 2018.
12. J. Jiang, V. Sekar, I. Stoica, and H. Zhang, "Unleashing the potential of data-driven networking," 2017.
13. C. Kim, A. Sivaraman, N. Katta, A. Bas, A. Dixit, L. J. Wobker, and B. Networks, "In-band network telemetry via programmable dataplanes," 2015.
14. D. Kreutz, F. M. V. Ramos, P. Esteves Verissimo, C. Esteve Rothenberg, S. Azodolmolky, and S. Uhlig, "Software-defined networking: A comprehensive survey," *Proceedings of the IEEE*, vol. 103, no. 1, pp. 10–13, 2014.
15. "Pnda." http://www.pnda.io/.
16. H. Mao, R. Netravali, and M. Alizadeh, "Neural adaptive video streaming with pensieve," in *Conference of the ACM Special Interest Group on Data Communication*, pp. 197–210, 2017.

17. C. Jiang, Y. Chen, Q. Wang, and K. J. R. Liu, “Data-driven auction mechanism design in iaas cloud computing,” *IEEE Transactions on Services Computing*, vol. PP, no. 99, pp. 1–1, 2015.
18. C. Qiu, S. Cui, H. Yao, F. Xu, F. R. Yu, and C. Zhao, “A novel qos-enabled load scheduling algorithm based on reinforcement learning in software-defined energy internet,” *Future Generation Computer Systems*.
19. J. A. Boyan and M. L. Littman, “Packet routing in dynamically changing networks: a reinforcement learning approach,” in *International Conference on Neural Information Processing Systems*, pp. 671–678, 1993.
20. S. C. Lin, I. F. Akyildiz, P. Wang, and M. Luo, “Qos-aware adaptive routing in multi-layer hierarchical software defined networks: A reinforcement learning approach,” in *IEEE International Conference on Services Computing*, pp. 25–33, 2016.
21. H. Zhang, B. Wang, C. Jiang, K. Long, A. Nallanathan, V. C. M. Leung, and H. V. Poor, “Energy efficient dynamic resource optimization in noma system,” *IEEE Transactions on Wireless Communications*, vol. PP, no. 99, pp. 1–1, 2018.
22. C. Fang, H. Yao, Z. Wang, P. Si, Y. Chen, X. Wang, and F. R. Yu, “Edge cache-based isp-cp collaboration scheme for content delivery services,” *IEEE Access*, vol. 7, pp. 5277–5284, 2019.
23. N. Mckeown, T. Anderson, H. Balakrishnan, G. Parulkar, L. Peterson, J. Rexford, S. Shenker, and J. Turner, “Openflow:enabling innovation in campus networks,” *Acm Sigcomm Computer Communication Review*, vol. 38, no. 2, pp. 69–74, 2008.
24. P. Bosshart, M. Izzard, M. Izzard, M. Izzard, N. Mckeown, J. Rexford, T. Dan, T. Dan, A. Vahdat, and G. Varghese, “P4: programming protocol-independent packet processors,” *Acm Sigcomm Computer Communication Review*, vol. 44, no. 3, pp. 87–95, 2013.
25. “barefootnetworks.” https://www.barefootnetworks.com/.
26. A. W. Moore and D. Zuev, “Internet traffic classification using bayesian analysis techniques,” *Acm Sigmetrics Performance Evaluation Review*, vol. 33, no. 1, pp. 50–60, 2005.
27. H. Zhang, L. Chen, B. Yi, K. Chen, Y. Geng, and Y. Geng, “Coda: Toward automatically identifying and scheduling coflows in the dark,” in *Conference on ACM SIGCOMM 2016 Conference*, pp. 160–173, 2016.
28. T. T. T. Nguyen and G. Armitage, “A survey of techniques for internet traffic classification using machine learning,” *IEEE Communications Surveys & Tutorials*, vol. 10, no. 4, pp. 56–76, 2009.
29. M. Yue, C. Jiang, T. Q. S. Quek, H. Zhu, and R. Yong, “Social learning based inference for crowdsensing in mobile social networks,” *IEEE Transactions on Mobile Computing*, vol. PP, no. 99, pp. 1–1, 2017.
30. C. Li and C. Yang, “The research on traffic sign recognition based on deep learning,” in *International Symposium on Communications and Information Technologies*, pp. 156–161, 2016.
31. K. Arulkumaran, M. P. Deisenroth, M. Brundage, and A. A. Bharath, “A brief survey of deep reinforcement learning,” 2017.
32. J. Schmidhuber, “Deep learning in neural networks: An overview.,” *Neural Networks the Official Journal of the International Neural Network Society*, vol. 61, p. 85, 2015.
33. M. Karakus and A. Durresi, “Quality of service (qos) in software defined networking (sdn): A survey,” *Journal of Network & Computer Applications*, vol. 80, pp. 200–218, 2017.
34. S. Levine, S. Levine, S. Levine, and S. Levine, “Continuous deep q-learning with model-based acceleration,” in *International Conference on International Conference on Machine Learning*, pp. 2829–2838, 2016.
35. “gym.” https://gym.openai.com/.
36. “keras.” https://keras.io/.
37. L. Cui, F. R. Yu, and Q. Yan, “When big data meets software-defined networking: Sdn for big data and big data for sdn,” vol. 30, pp. 58–65, 2016.
38. S. J. Pan and Q. Yang, “A survey on transfer learning,” *IEEE Transactions on Knowledge & Data Engineering*, vol. 22, no. 10, pp. 1345–1359, 2010.
39. T. Huang, F. R. Yu, C. Zhang, J. Liu, J. Zhang, and J. Liu, “A survey on large-scale software defined networking (sdn) testbeds: Approaches and challenges,” *IEEE Communications Surveys & Tutorials*, vol. PP, no. 99, pp. 1–1, 2017.

Chapter 3
Intelligent Network Awareness

In the network, different applications produce various traffic types with diverse features and service requirements. Therefore, in order to better manage and control networking, the intelligent awareness of network traffic plays a significant role. Network traffic information mainly includes service-level information (e.g., QoS/QoE), anomaly traffic detection information, etc. In this chapter, we first present a multi-level intrusion detection model framework named MSML to address these issues. The MSML framework includes four modules: pure cluster extraction, pattern discovery, fine-grained classification and model updating. Then, we propose a novel IDS framework called HMLD to address these issues, which is an exquisitely designed framework based on Hybrid Multi-Level Data Mining. In addition, we propose a new model based on big data analysis, which can avoid the influence brought by adjustment of network traffic distribution, increase detection accuracy and reduce the false negative rate. Finally, we propose an end-to-end IoT traffic classification method relying on deep learning aided capsule network for the sake of forming an efficient classification mechanism that integrates feature extraction, feature selection and classification model. Our proposed traffic classification method beneficially eliminates the process of manually selecting traffic features, and is particularly applicable to smart city scenarios.

3.1 Intrusion Detection System Based on Multi-Level Semi-Supervised Machine Learning

With the rapid development of the Internet, the number of network invasions greatly increases [1]. As a widely-used precautionary measure, intrusion detection has become an important research topic. Machine learning (ML), which can address many nonlinear problems well, has gradually become the mainstream in the field of

H. Yao et al., *Developing Networks using Artificial Intelligence*, Wireless Networks, https://doi.org/10.1007/978-3-030-15028-0_3

intrusion detection system. Many existing models based on ML employ supervised learning algorithms to train an intrusion detection classifier using a set of labeled training samples, then use this classifier to classify unlabeled test data.

There are two main problems in network traffic, which affect the robustness of the ML model. Firstly, network traffic has a severe class-imbalance problem. It means that some categories have much more samples than others. This problem is also called 'elephant traffic' and 'mouse traffic' problem. The generated ML model will suffer because the model will be much more suitable for 'elephant traffic' rather than 'mouse traffic'. Secondly, training data and test data can have a non-identical distribution problem. It means that our training data and test data are generated by two different probabilistic distributions [16, 31]. The independent identical distribution is an extremely important premise of statistical machine learning. Non-identical distribution problem can cause a decrease in accuracy rate. Unfortunately, the distribution of network traffic is not static because the uncertainty of users' behaviors leads to the variable distribution of network traffic. The high cost of expert system determines that it is not possible to mark a large amount of network traffic in real-time [39]. We usually train the model using historical labeled data.

In this section, we propose a novel intrusion detection system based on a multi-level semi-supervised machine learning framework (MSML) to tackle with aforementioned problems. The main contributions of this part are described as follows:

- In order to alleviate the class imbalance problem, we propose a concept of pure cluster, and propose a cluster-based under-sampling methods, which is a hierarchical, semi-supervised k-means clustering algorithm. There exists a large number of samples in pure cluster. Furthermore, these pure cluster samples can be predicted accurately using our proposed method.
- For the non-identical distribution problem, a method is proposed to distinguish known and unknown pattern samples in the test set. We can obtain a guaranteed high recognition accuracy for known pattern samples. Also, the unknown pattern samples could be fine-grained classified.

In this part, the MSML framework is experimented and evaluated on the KDD-CUP99 dataset. The results show the proposed framework significantly outperforms our baseline method and many other existing models in the respect of overall accuracy, F1 score and unknown pattern discovery ability. The results also show that when the non-identical distribution problem do not occur, our framework is still suitable and well-performed.

3.1.1 *Proposed Scheme (MSML)*

In this subsection, we introduce our proposed MSML framework, as shown in Fig. 3.1. The data generator process for MSML is also shown in Fig. 3.1.

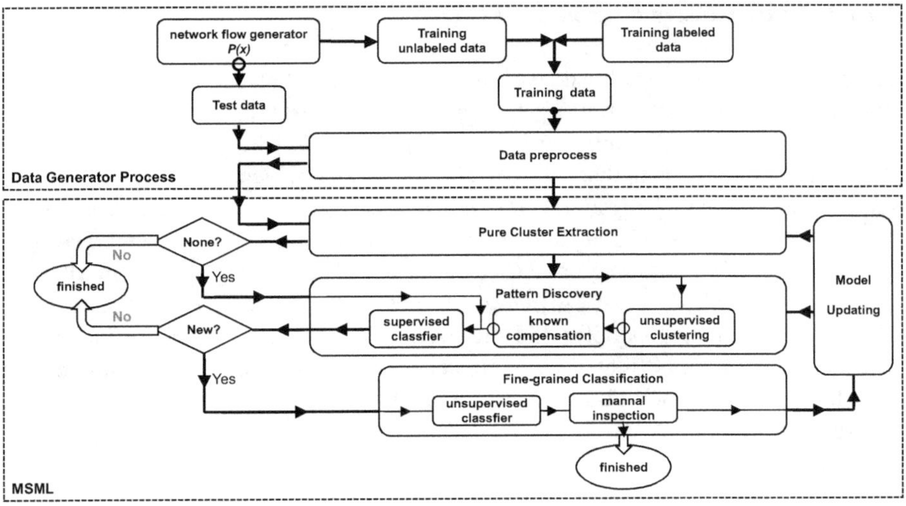

Fig. 3.1 The MSML framework

The aim for data generator process is to generate the required training set and test set for the MSML framework. Due to the semi-supervised property of the MSML, the training set consists of labeled samples and unlabeled samples. The training labeled data was labeled in the past, reflecting the distribution of historical known network traffic. The training unlabeled samples and test samples are all generated by the network traffic generator, which reflects the distribution of current network traffic. The data preprocessing module is devised to do something necessary before training model such as normalization and data cleaning.

The MSML consists of four modules including pure cluster extraction, pattern discovery, fine-grained classification and model updating. The pure cluster extraction module aims to find large and pure clusters. In the pure cluster module, this section defines an important concept of "pure cluster pattern" and proposes a hierarchical semi-supervised k-means algorithm (HSK-means), aiming at finding out all the pure clusters. In the pattern discovery module, this section defined the "unknown pattern" and applied cluster based method to find those unknown patterns. The fine-grained classification module achieves fine-grained classification for those unknown pattern samples. The model updating module provides a mechanism for retraining. For any test sample, once labeled by one module, will not go on any more; and all test samples will be labeled in pure cluster extraction module, pattern discovery module and fine-grained classification module.

3.1.1.1 Pure Cluster Extraction (PCE)

The pure cluster can be defined as a cluster where almost all samples have the same category. The potential samples located in this pure cluster can be considered to be the same category as the other samples in this cluster [33].

Given a training labeled set, $D^l = \{l_1, l_2, \cdots, l_N\}$, which consists of N labeled samples and a unlabeled training set, $D^u = \{u_1, u_2, \cdots, u_M\}$, which consists of M unlabeled samples. The labeled and unlabeled training samples are merged with an aim to form our training set $D = D^l \cup D^u$. The training set are partitioned into K clusters $\{C_1, C_2, \ldots, C_K\}$ $(K \leq |D|)$ using the K-means clustering method. If the training labeled samples and the training unlabeled samples are identically distributed, we assume that a large cluster C_i which contains a large number of samples will meet the following formula, by the principle of the central limit theorem.

$$\frac{|C_i^l|}{|C_i|} \approx \frac{N}{N+M}, \tag{3.1}$$

where C_i^l is the k-th cluster in the labeled samples. We can conclude that the cluster is not pure, if the left hand side of Eq. (3.1) is significantly smaller than the right hand side. The reason is that the labeled samples only characterize a part of the whole cluster. Consequently, we mark a cluster "pure" cluster when all the labeled samples of this cluster belong to a same category and the cluster meets the following formula:

$$\frac{|C_i^l|}{|C_i|} \geq \eta * \frac{N}{N+M}, \tag{3.2}$$

$$|C_i| \geq MinPC, \tag{3.3}$$

where $MinPC$ is a parameter reflecting the minimum number of samples of a cluster which is required to be spilt using a cluster method recursively. η is a real number and cannot exceed 1. However, the value of η cannot set to be too small. We proposed a hierarchical semi-supervised K-means (HSK-means) based on the above mentioned definition of pure cluster. The pseudo-codes of training phases of HSK-means algorithm are shown in Algorithm 1. We define a new parameter $ArsPC$, reflecting the average number of samples of clusters when HSK-means employ the clustering in all the samples of a parent cluster. The formula (3.4) shows the method of calculating the number of child clusters. The key point is that if a cluster do not meet formula (3.2), then we will perform the K-cluster algorithm on this cluster recursively unless this cluster do not meet formula (3.3), as is shown in the 10th line in Algorithm 1. The pseudo-codes of training phases of HSK-means algorithm are shown in Algorithm 2.

$$K = \frac{|C|}{ArsPC}. \tag{3.4}$$

Algorithm 1 MSML-PCE HSK-means training

```
Require: training set D
Ensure: new training set D*
function TREECLUSTER(D)
    Calculate the number of clusters using formula (3.4)
    Perform clustering on D to obtain clusters {C_1, C_2, ... C_K}
    D.childern ← {C_1, C_2, ... C_K}
    for i = 1 → K do
        if the cluster C_i is a pure cluster then
            Label all samples in C_i
            Update D* with D − C_i
        else if |C_i| ≥ ArsPC then
            TreeCluster(C_i)
        end if
    end for
end function
```

Algorithm 2 MSML-PCE HSK-means test

```
Require: test set T
Ensure: new test set T*, labeled test set T_finished
for ∀s_i ∈ T do
    index ← D
    while index has children do
        Find the nearest children C_j from s_i
        index ← C_j
        if index has a label C then
            Label s_i with C
            Put s_i into Te_finished
            break
        end if
    end while
    if s_i has not a label then
        Put s_i into T*
    end if
end for
```

The time complexity of the HSK-means algorithm is $O(ntk)$, where n is the number of all the samples in the training set, t is the number of iterations with the general k-means, k is the number of all the clusters including those generated by their parent clusters and those generated by the whole data. The HSK-means algorithm is faster than general k-means when both of clusters are identical. In the HSK-means algorithm, some of clusters is generated by their parent clusters, the number samples of which is obviously much less than the whole data [36].

For the training set, any sample in pure clusters can be extracted from the training set. We only preserve those non-pure cluster samples for the training set of the next module. This is actually a under-sampling method based on cluster. For those pure clusters, the sampling rate is zero. For those non-pure clusters, the sampling rate is one. For the test set, all samples in the pure clusters are labeled while all samples in those non-pure cluster are not labeled and are preserved for the test set of the next module.

If all pure clusters cover overmuch samples, the module will leave little remaining training data to next module, thereby falling into overfitting. Hence, the appropriate values of two parameters $ArsPC$ and $MinPC$ are important for the whole MSML framework. It is necessary to adjust these values of the two parameters.

3.1.1.2 Pattern Discovery (PD)

Pattern is an abstract concept. In this chapter, it refers to a kind of data distribution in the feature space. Pattern can be classified in different ways [40]. For example, pattern can be divided into known pattern and unknown pattern. The known patterns refer to all patterns that exist in the training set, while the unknown patterns refer to patterns that do not exist in the training set. For another example, pattern can also be classified into intrusion pattern and normal pattern. The intrusion patterns refer to patterns of labeled training intrusion samples, while the normal patterns refer to patterns of labeled training normal samples. In a word, pattern is an abstract description about data distribution that we wish to recognize. What we wish to recognize determines the definition of specified pattern. Then we can apply heuristic and machine learning methods to recognize it. For example, the pure cluster introduced in Sect. 3.1.1 are belonged to a kind of pattern. Then we use our proposed hierarchal semi-supervised K-means algorithms to obtain the required pure clusters.

In this chapter, the main criteria that determines a sample to be known or unknown is the pattern of this sample rather than the sample category. Our hypothesis about the real distribution of network traffic is that the majority of known traffic samples come from known patterns while the majority of unknown traffic samples come from unknown patterns. That means, the probability that a known traffic sample is an unknown pattern sample is very close to zero. It is so distinctive to the work in [7, 9] where the criteria is traffic category. From the point of view of patterns, [6, 8, 9] indicated an assumption that unknown traffic was unknown pattern and known traffic was known pattern.

This module has three sub-modules, clustering, known compensation and supervised classifier. The pseudo-codes of training and test phases of the pattern recovery module are shown in Algorithm 3. All training samples which are not located in pure clusters form this module's training set. Similarly, all test samples which are not labeled in previous module form this module's test set.

We applied a semi-supervised algorithm named K-means in the process of clustering. K-means employ a clustering in the whole training set Tr_{pd}. We define a new parameter AsPD to denote the average size of clusters . Obviously, the number of clusters for the clustering can be expressed by $K = |Tr_{pd}| \ / \ AsPD$. If the training labeled samples and the training unlabeled samples are identically distributed, a large cluster, which contains a large number of samples, will meet

Eq. (3.1) by the central limit theorem. However, if a cluster does not meet Eq. (3.1) but it meets the following formula, we regard the cluster as a unknown pattern and label all the samples in the cluster as "new".

$$\frac{|C_i^l|}{|C_i|} < \mu * \frac{N}{M} \tag{3.5}$$

$$|C_i| \geq AsPD \tag{3.6}$$

where μ is a real number. Its value is suggested to a small number.

The second step is called known compensation. With the decrease of the value of AsPD, both TP (true positive rate) and FP (false positive rate) increase. The reason is that some known patterns are considered as unknown patterns. This can significantly reduce the precision of the unknown patterns and can increase the complexity of the internal structure of the unknown patterns. In order to maximize the increase of TP while minimizing the increase of FP, we make some compensation. The idea is to use other supervised learning algorithms to train the labeled data and calculate the confidence of each unknown patterns. Those patterns whose confidence are large enough will not be considered as unknown patterns any more. We define a confidence minimum threshold. In this chapter, we use the softmax regression algorithm to calculate this confidence. Considering the softmax regression is sensible to unbalanced data, this section presents a cluster-based under-sampling technique applying a under-sampling rate function. The larger the cluster is, the value of the under-sampling rate function is.

The third step is "supervised classifier". After the second step, the training set is composed of one normal category, N intrusion categories, and one unknown category. The normal and intrusion categories are all regarded as known patterns. The unknown category represents all the unknown patterns. In this chapter, we use a random forest algorithm with an aim to build a supervised classifier. After this step, all the test samples are labeled. We successfully separate all the test samples into known patterns and unknown patterns. If unknown patterns are essential to perform a fine-grained classification, then next model will work.

The time complexity of Algorithm 3 depends on the time complexities of its three sub algorithms, including K-means, logistic regression and random forest. Its time complexity equals to the sum of the time complexities of the three sub algorithms.

3.1.1.3 Fine-Grained Classification (FC)

After the processing of pattern discovery module, all the test samples are labeled. However, some samples are labeled as "new", which is neither a normal category nor an intrusion category, but a new category. Expert inspection is used to achieve fine-grained classification. We have high confidence to classify these samples correctly with low artificial cost because we have separated a few of unknown patterns from a group of complicated patterns. Algorithm 4 is the pseudo-code of fine-grained

Algorithm 3 MSML-PD

Require: training set D^*, test set T^*, labeled test set $T_{finished}$
Ensure: labeled test set $T_{finished}$, unknown test set T'
$T' \leftarrow \{\}, \quad D_s^* \leftarrow \{\}, \quad D_n \leftarrow \{\}$
Perform clustering on D^* to obtain clusters $\{C_1, C_2, \ldots C_K\}$
for $i = 1 \rightarrow K$ **do**
  if C_i is an unknown pattern **then**
    $D_n \leftarrow D_n \bigcup C_i$
  end if
  Generate $C_i^s \subseteq C_i$
  Put q into D_s^*, $\forall q \in C_i^s$
end for
Choose training labeled samples D_l^* from D_s^*
Train a softmax classifier using D_l^*
for all $C_i \in D_n$ **do**
  if $\exists$ $category\ c$ & minimum probability of $c \geq \alpha$ **then**
    $D_n \leftarrow D_n - C_i$
  end if
end for
Choose training labeled samples D_l from D^*
Combine D_l with D_n to train a supervised classifier f
for $\forall s \in T^*$ **do**
  if s is classified as new by classifier f **then**
    Put s into T'
  elsePut s into $T_{finished}$
  end if
end for

Algorithm 4 MSML-FC

Require: labeled test set $T_{finished}$, unknown test set T'
Ensure: labeled test set $T_{finished}$ which includes some additional samples
Perform clustering on T' to obtain clusters $\{C_1, C_2, \ldots C_K\}$
for $i = 1 \rightarrow K$ **do**
  $count \leftarrow 0$
  while $count < MaxFC$ **do**
    Randomly select A samples from C_i
    Expert inspection
    if all the A samples has the same ground-truth category C **then**
      Label all samples in C_i with C;
      $break$;
    end if
  end while
  if none sample in C_i is labeled **then**
    Label all samples in C_i with $suspious$
  end if
  Put all samples in C_i into $T_{finished}$
end for

classification module. we perform k-means clustering on these unknown pattern samples first. For each cluster, we randomly select several samples (e.g., three samples) for manual inspection. If all the selected samples have the same ground-truth category, then we achieve fine-grained classification, as is shown in 5–9 lines in Algorithm 3. The parameter $MaxFC$ indicates the number of allowed attempts.

The time complexity of Algorithm 4 depends on the time complexity of the K-means and the velocity of expert inspection process.

3.1.1.4 Model Updating

In this section we continue to discuss how we can update our model. When the amount of "new" samples is relatively large, it is possible to train a supervise model based on the samples of these unknown patterns. In this manner, a consecutive sample of the "new" can be identified as a specific class directly by this model. Only the current distribution of network traffic generator does not change, it is effective to do so.

When the distribution varies due to going through a long period of time, introducing a feedback mechanism is necessary. If a new cluster is pure and have enough samples and the cluster does not overlap with other prepared pure clusters in feature space, then it is time to update the pure cluster extraction module. Otherwise we should update the pattern discovery module. In doing so, the model can be always able to adapt to the new traffic distribution.

3.1.1.5 The Hyper-Parameters

The main hyper-parameters of MSML are listed as shown in Table 3.1.

Table 3.1 Main hyper-parameters

Name	Module	Definition
`ArsPC`	MSML-PCE	The average number of samples of clusters
`MinPC`	MSML-PCE	The minimum number of samples of a cluster which is required to be spilt using a cluster method recursively
`ArsPD`	MSML-PD	The average number of samples of clusters
`MaxFC`	MSML-FC	The number of attempts to conduct inspection allowed

3.1.2 *Evaluation*

3.1.2.1 Dataset

We choose the KDDCUP99 dataset to evaluate the MSML framework [43]. The KDDCUP99 dataset contains four intrusion categories and one normal category. The four intrusion categories are DOS, R2L, U2R and Probe, respectively. Each of intrusion categories contains several of subcategories. The KDDCUP99 training dataset contains 5 categories and 22 subcategories. The KDDCUP99 test dataset contains 17 subcategories of 4 categories which do not appear in the training dataset. The reason why we choose KDDCup99 as our dataset consists of two aspects. The one is that the source of dataset we can get is limited, the other one is that there are a large number of research works using this dataset can be compared with our proposed method.

Inconsistent Dataset
In order to evaluate the performance of our MSML framework on non-identical distribution dataset, we construct a dataset named *inconsistent dataset*. The training dataset is composed of two parts including labeled samples and unlabeled samples. We choose a fraction of the KDDCUP99 training dataset as labeled samples due to the fact that using all of the KDDCUP99 training dataset is time-consuming. We randomly select 20% of the KDDCUP99 test dataset as unlabeled samples due to the fact that the training unlabeled samples and test samples need to be identically distributed according to our MSML framework. Table 3.1 shows the composition of the compositive training set of inconsistent dataset (Table 3.2).

Consistent Dataset
In order to evaulate the performance of our MSML framework on identical distribution dataset, we construct a dataset named *consistent dataset*. We divide the KDDCUP99 training dataset into three subsets according to the rate of 20%, 20% and 60%, respectively. These three subsets represent training labeled samples, training unlabeled samples and test samples, respectively.

Table 3.2 Composition of training set

Category	Number
Normal	3242
DOS	1026
Probe	7829
U2R	52
R2L	563
Unlabeled	60,020

3.1.2.2 Data Pre-process

The KDDCUP99 dataset contains 41 features including nine discrete features and 32 consecutive features. We adopt one-hot and numerical-order methods to deal with discrete features. We employ one-hot when using ANN and K-means due to the fact that numerical-order can bring nonexistent sequence influence. We adopt numeric-order when using other methods because one-hot can greatly increase the data sparsity.

In this chapter, 32 consecutive features are normalized. However, the values of some consecutive features (“duration”, “src_bytes”, “num_root”, “num_compromised”, “_bytes”) show unusual value distribution. For these unusual features, the majority of the values are much smaller than the maximum value, which means 0–1 normalization will make majority of values close to zero. To avoid this situation, we adopt logarithmic normalization processing. Logarithmic normalization does not change the order of the values, but it can significantly reduce the effect of abnormal maximum value.

3.1.2.3 Evaluation Criteria

The TP, TN, FP, FN[10] are usually used to evaluate the performance of machine learning model, which can be described by a confusion matrix shown in Table 3.3.

The Precision, Recall, F1_score and Accuracy [10] are also defined to evaluate the model performance.

$$P = Precision = \frac{TP}{TP + FP} \tag{3.7}$$

$$R = Recall = \frac{TP}{TP + FN} \tag{3.8}$$

$$F1\_score = \frac{2 \times Precision \times Recall}{Precision + Recall} \tag{3.9}$$

$$Accurary = \frac{TP + TN}{TP + TN + FN + FP} \tag{3.10}$$

In addition, we define several of evaluation indexes for MSML framework including *Capture rate*, *Coverage rate* and *Coverage capture rate*. After pattern

Table 3.3 Confusion matrix

	Predicted	
Actual	Positive	Negative
Positive	TP	FN
Negative	FP	TN

Table 3.4 Self-defined indexes

Index name	*Capture rate*	*Coverage rate*	*Coverage capture rate*
Index value	*B/M*	*b/M*	*b/B*

Table 3.5 Confusion matrix(baseline, inconsistent dataset)

	Predicted				
Actual	Normal	DoS	R2L	U2R	Probe
Normal	60,255	743	23	13	259
DoS	6203	223,364	8	1	277
R2L	15,627	0	539	17	6
U2R	134	0	8	30	56
Probe	620	154	2	0	3390

discovery module, *Capture rate* is the proportion of all test samples which are labeled. *Coverage rate* is the proportion of all test samples which are correctly labeled. *Coverage capture rate* is the proportion of test labeled samples which are correctly labeled. We suppose the test set has M samples. After performing the operation of pattern discovery module, we classify B samples, therein, we correctly classify b samples. Apparently, the remaining $M - B$ samples are labeled "new", waiting for being fine-grained labeled. The calculation ways of three indexes are shown in Table 3.4.

3.1.2.4 Baseline Model

In order to validate the effectiveness of the MSML framework, we adopt traditional 1+N model as our compared baseline model. Our experiment based on the baseline model is carried out on both inconsistent dataset and consistent dataset.

Inconsistent Dataset Comparison
We adopt the labeled samples from the training set of inconsistent dataset to train the baseline model, and evaluate the baseline model on the test set of inconsistent dataset. We take use of several supervised learning algorithms, including Naive Bayes, BP neural network, random forests, support vector machines and so on. Experimental results indicate that random forest achieves the highest overall accuracy by 92.46%. Table 3.5 shows the confusion matrix on the test set.

Furthermore, through the deep analysis of Fig. 3.2, we can conclude that the overall accuracy of the known traffic samples can reach 97.94%. However, the accuracy of unknown traffic samples is only 6.87%. The unknown traffic samples cause a decrease in not only overall accuracy, but also recall rate of all categories, as shown in Fig. 3.2.

Consistent Dataset Comparison
An experiment with the similar method as on the inconsistent dataset is conducted on the consistent dataset. We adopt the labeled samples of the consistent training

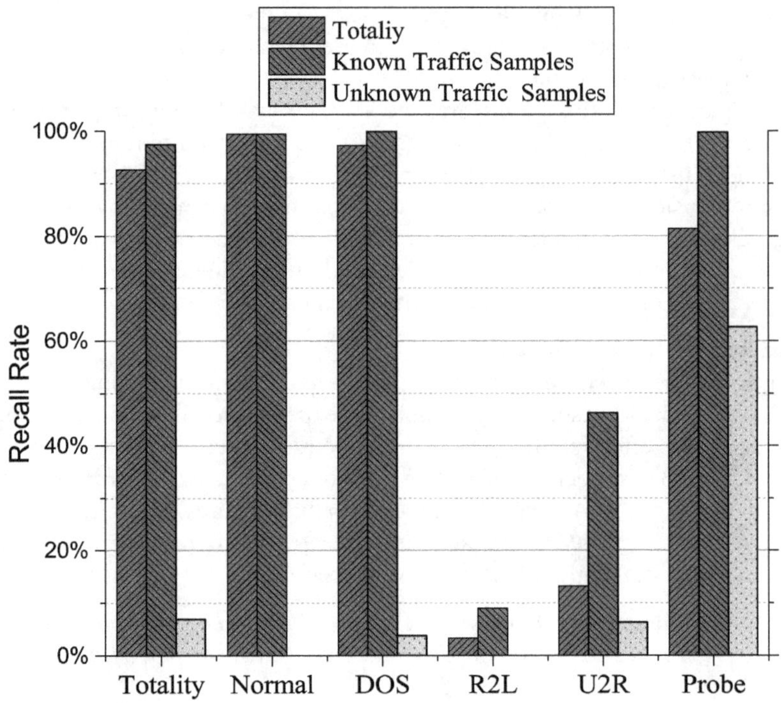

Fig. 3.2 Recall rate comparison of each category in baseline model

dataset to train our model. We employ this trained model to evaluate on two test set. One is the whole test set of the consistent dataset. We obtain 99.92% overall accuracy. 99.92%, which is so close to 100%, representing the high recognition capability of known pattern samples. The other is the unknown traffic portion of test set of the inconsistent dataset. We obtain 97.9% accuracy, worse than the value of 99.92%. We attribute the difference to non-identical distribution. Some known traffic samples in the test set of inconsistent dataset actually belong to unknown patterns thereby deteriorating the overall accuracy.

3.1.2.5 MSML

In addition to the overall accuracy, this part claims that the *capture rate* and *coverage capture rate* are also significant. We make the index of *coverage capture rate* have a high value in order to reduce the influence of error classification on intrusion detection. The two indexes of *coverage capture rate* and *capture rate* are often contradictory, due to the fact that the model is apt to consider a portion of known pattern samples as unknown pattern samples. In this way, it can lead to a decrease for the index of capture rate and an increase for the index of coverage capture rate.

Meanwhile, it can greatly increase the complexity of the structure of internal feature space for unknown pattern samples and increase the burden on the subsequent fine-grained classification module.

In this part, there are several parameters which may have an important impact on the above two indexes of coverage capture rate and capture rate. For example, the average cluster size of pure cluster extraction module denoted by $ArsPC$, the lower limit cluster size of pure cluster extraction module denoted by $MinPC$, and average cluster size of pattern discovery module denoted by $AsPD$, these three parameters are of great concern to us. We set different values to these parameters in our experiments on both inconsistent dataset and consistent dataset.

Inconsistent Dataset Comparison

We conduct some experiments on the inconsistent dataset. The parameter $AsPD$ is set to 20, 50, and 100, respectively. Simulation experiments show that the parameter values have little effect on the results. In our work, we set $AsPD$ to value of 100. Figures 3.3 and 3.4 respectively show the trend of *coverage capture rate* and *capture rate* with the change of $ArsPC$ and $MinPC$. In these figures, each circle represents a value of *coverage capture rate* and *capture rate*, the more red the circle is, the bigger the circle is, the higher the circle value becomes.

Figure 3.3 reflects the relationship between *coverage capture rate* and $ArsPC$, $MinPC$. From Fig. 3.3, we can observe that the *coverage capture rate* is generally

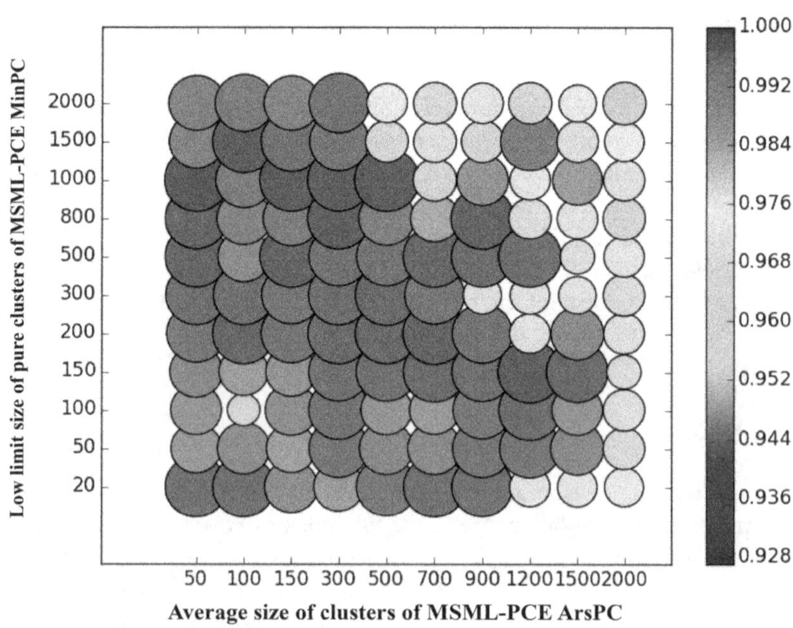

Fig. 3.3 MSML: coverage capture rate

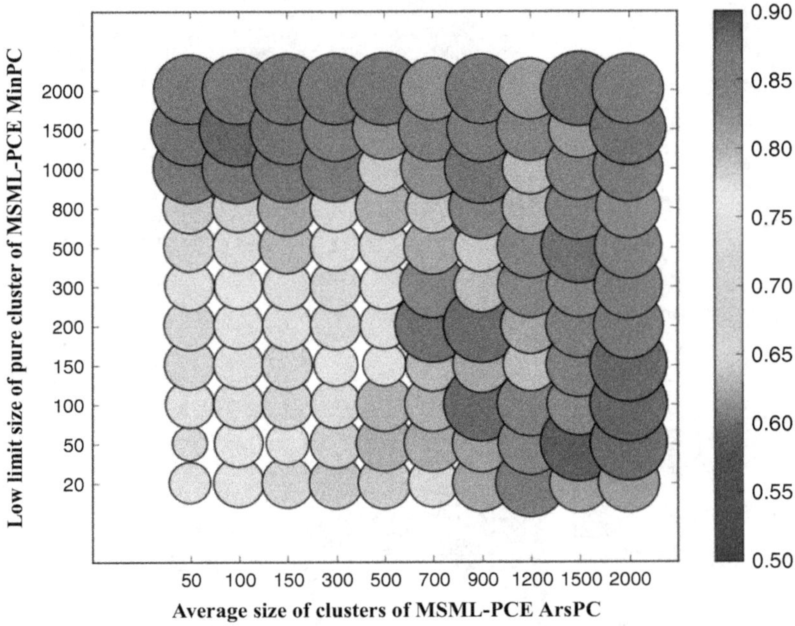

Fig. 3.4 MSML: capture rate

97% at least. In addition, with the increase of $ArsPC$ and $MinPC$, the *coverage capture rate* has a tendency of sightly increase first then significantly decrease.

Figure 3.4 reflects the relationship between *capture rate* and $ArsPC$ and $MinPC$. From Fig. 3.4, we can observe that the capture rate is relatively low when $ArsPC$ and $MinPC$ have lower values. In addition, the overall trend of *capture rate* is increasing with the increase of $ArsPC$ and $MinPC$. Furthermore, there is a certain randomness when the values of $ArsPC$ and $MinPC$ are large, therefore, when the values of $ArsPC$ and $MinPC$ are bigger than a predefined threshold, it is not necessary to continue to increase their values.

Considering both *coverage capture rate* and *capture rate*, it is proper to choose a lower $ArsPC$ and a larger $MinPC$. In this chapter, the parameter $ArsPC$ is set to 100 and the parameter $MinPC$ is set to 1500. The accuracy of MSML can reach 96.6%, which has greatly improved compared with the baseline model, which can reach 92.5%. It can be seen in Table 3.6.

In the fine-grained classification module, 150 samples, whose proportion is less than 0.05%, are conducted expert inspection. The result of expert inspection is then applied to all the unknown pattern samples, whose proportion is about 12%. The accuracy of unknown pattern samples is 76.7%.

With respect to the recall rate, as is shown in Fig. 3.5, the elephant traffic DOS has improved to some extent, while the common traffic has a great improvement. For the mouse traffic, U2R and R2L have particularly notable improvement. The

Table 3.6 Comparison

Method	Normal	DoS	Probe	R2L	U2R	Accuracy
SVM+ELM+MK[13]	98.1	99.5	87.2	21.93	31.39	95.79
NFC[11]	98.2	99.5	84.1	14.1	31.5	N/A
SVM+BIRCH[35]	99.3	99.5	97.5	19.7	28.8	95.7
MOGFIDS[2]	98.4	97.2	88.6	15.8	11.0	93.2
Association rules[12]	**99.5**	96.6	74.8	3.8	1.2	92.4
Baseline	99.4	97.2	81.4	3.3	13.2	92.5
MSML	86.2	**99.8**	**98.4**	**90.6**	**72.5**	**96.6**

Bold values indicate the maximum value of each column

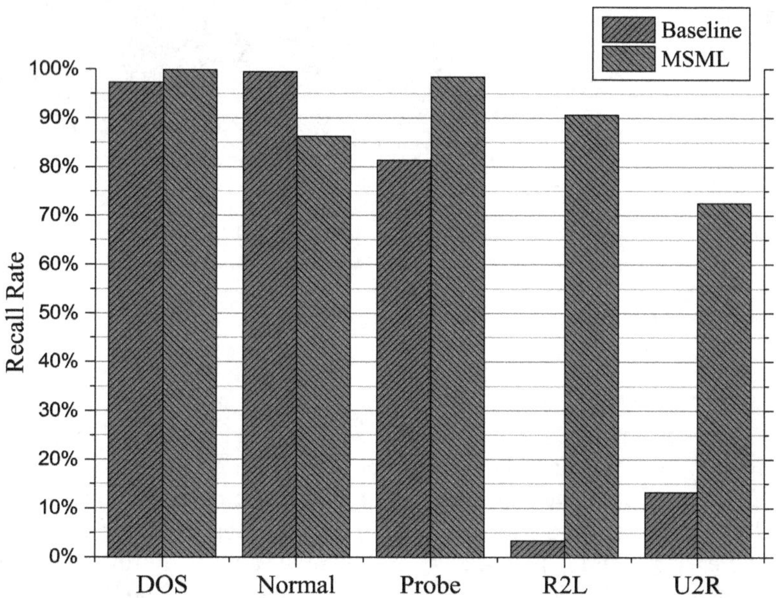

Fig. 3.5 MSML: recall

recall rate of U2R and R2L have greatly increased from 13.2% to 72.5%, and 3.3% to 90.6%, respectively. With respect to the precision, as is shown in Fig. 3.6, the precision of DOS remains 99.9%, and the precision of Normal, U2R, and R2L have improved at different degree. With respect to the F1 score, as is shown in Fig. 3.7, the F1 score of all categories improves. The F1 score of Normal, Dos, Probe, R2L and U2R have increased from 0.885 to 0.912, from 0.985 to 0.999, from 0.832 to 0.934, from 0.064 to 0.754 and from 0.208 to 0.743, respectively.

However, we must note that the recall rate of normal and the precision rate of R2L have a descending trend. This situation should be further investigated. Another experiment is conducted. In the model updating module, we find a suspicious cluster, where mixed snmpgetattack and snmpguess of R2L with Normal samples. The number of R2L samples and the number of normal samples is about 53 to 47.

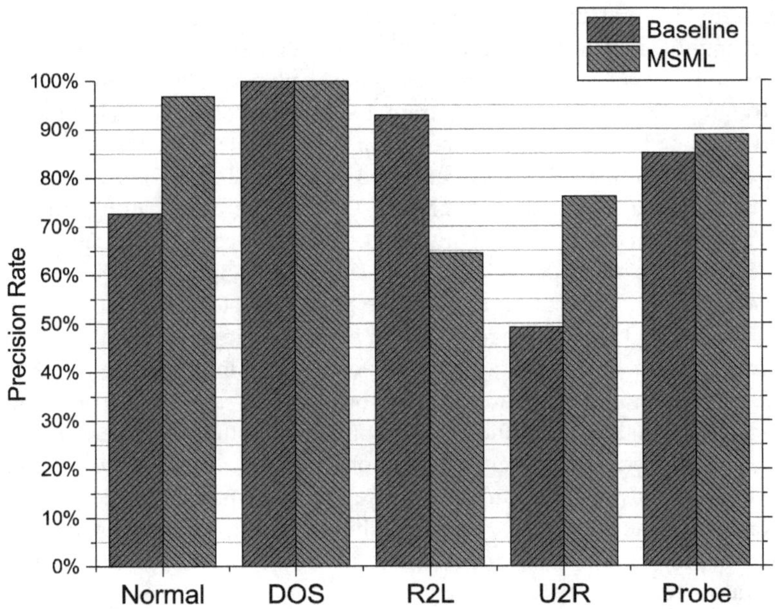

Fig. 3.6 MSML: precision

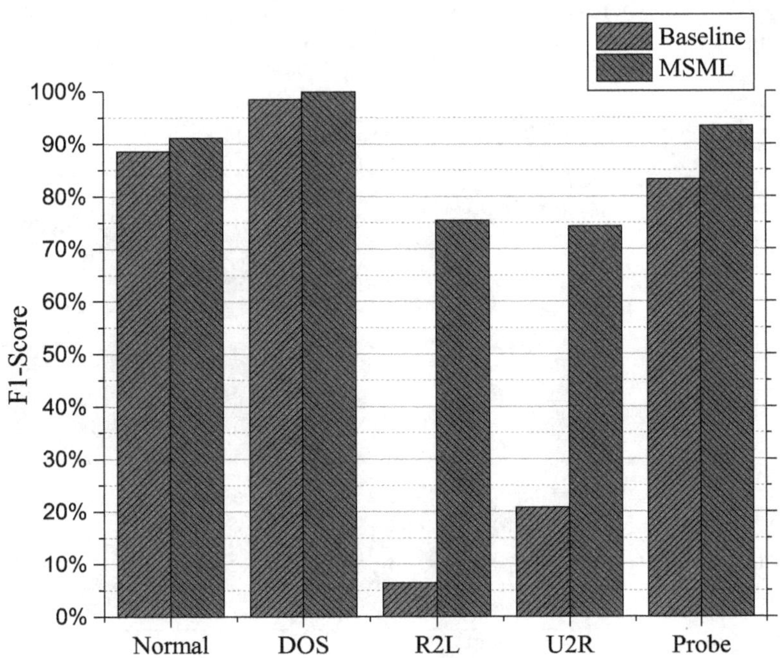

Fig. 3.7 MSML: F1-score

Further study[13] find that, in this cluster, samples of R2L and Normal in the feature space are highly analogous so it is difficult to expect them distinguished. To confirm this conjecture, we make all the samples of the suspicious cluster randomly divided into a training set and a test set. We train a supervised classifier using the training set, and evaluate on the test set. Experimental result show that the accuracy of the test set is no higher than 53%. It is illustrated that samples of R2L and normal really cannot be separated in this kind of cluster. We take the principle of priority of the invasion. Hence, all the samples in this cluster are sentenced to R2L. This is different from the baseline experiment, where all of samples are adjudged to Normal. It is the reason why the recall rate of normal and the precision rate of R2L decrease.

Figures 3.8 and 3.9 show the relationship between categories and patterns. Figure 3.8 shows the ratio of known traffic to unknown traffic in the whole test dataset, in the unknown pattern samples, and in the known pattern samples respectively. It can be seen that the unknown traffic ratio of unknown pattern samples to the whole test dataset has a great increment. We can observe that there are 89% of unknown traffic which is considered as unknown pattern. Meanwhile, 5% of known traffic is considered as unknown pattern. Figure 3.9 shows the ratio of known pattern samples to unknown pattern samples in the whole test set, in the unknown traffic, and in the known traffic, respectively.

We conduct two experiments with an aim to verify that the HSK-means algorithm in the pure cluster extraction module is important and indispensible in our MSML framework. One experiment begins with the pattern discovery module lacking of the pure cluster extraction module. The result is shown as Fig. 3.10. From Fig. 3.10 we

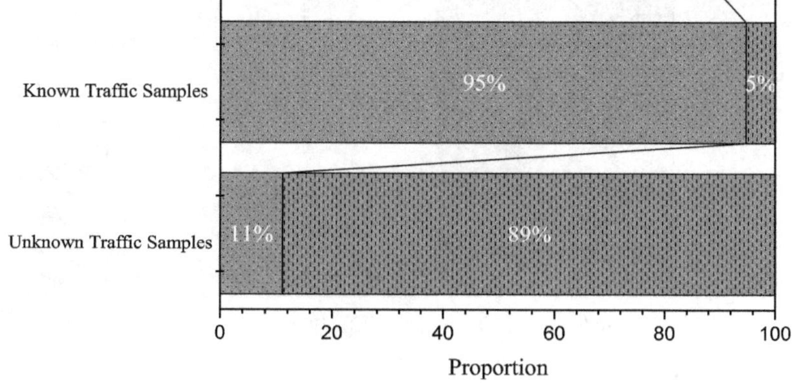

Fig. 3.8 Pattern in traffic

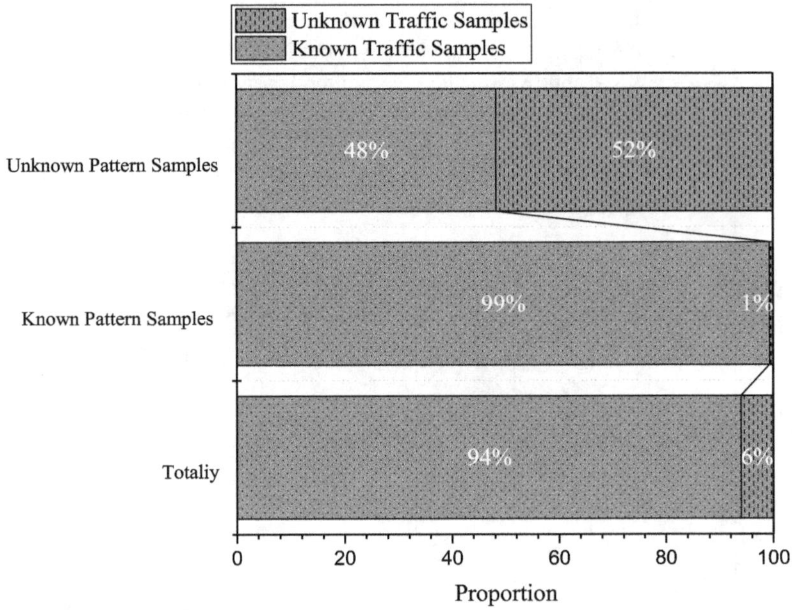

Fig. 3.9 Traffic in pattern

can see that the capture rate decline and coverage capture rate rise with the increase of the number of the cluster in pattern discovery module. However, when the coverage capture reaches more than 99% as we anticipate, the capture rate is about 78%, which is much smaller than the counterpart which can reach about 88% of the MSML framework. The other experiment replaces the HSK-means algorithm with a common under-sampling method. The result, as is shown in Fig. 3.10, demonstrated that the coverage rate cannot reach 96.7% in any case. We can conclude that the pure cluster extraction is important and indispensible to our MSML framework based on the results of these two experiments.

The performance comparison of MSML and other models is illustrated in Table 3.6. MOGFIDS and baseline models are the most common "1+N" supervised learning models. Both of them have a good detection rate in Normal and DOS, but the detection rate of U2R and R2L is very bad. Furthermore, the overall accuracy is also at a low level. Association rules[12] is a pure unsupervised learning algorithm combining heuristic rules that can hardly be identified by the rat categories. Both of [13] and [35] apply ways to identify unknown samples. Their detection rate of DOS is particularly high, and the detection rate of other categories is also at a high level. For our MSML, the detection rate of DOS, Probe, U2R and R2L and the overall accuracy are the highest. We has been analyzed that it is the defects of KDDCUP99 test set that lead to a decrease for the detection rate of Normal in MSML. Therefore, we can make the conclusion that MSML-IDS has a strong robustness.

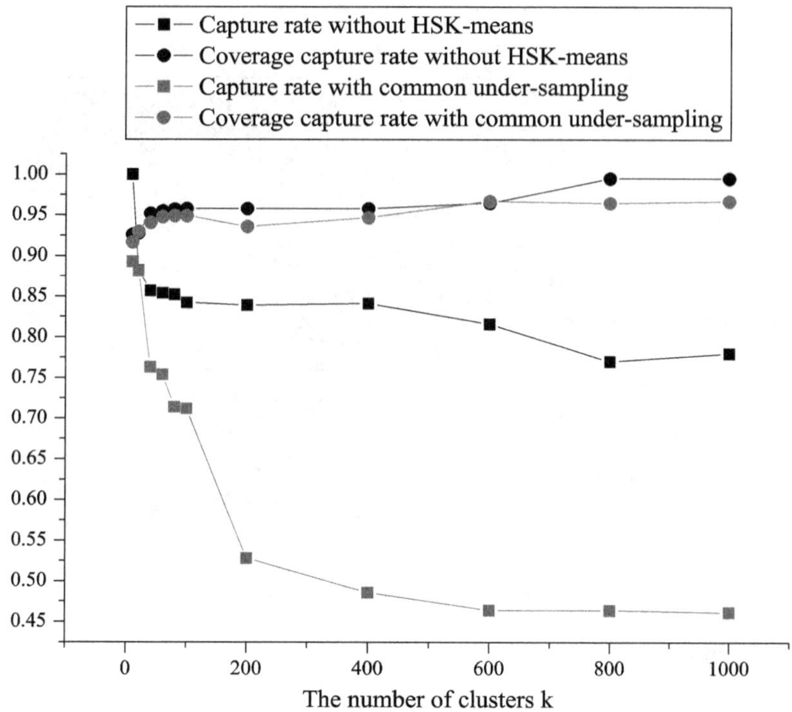

Fig. 3.10 The importance of HSK-means

Consistent Dataset

The framework of MSML is also employed in the consistent dataset. In this dataset, whatever the values of $ArsPC$, $MinPC$, and $AsPD$ are, we obtain a capture rate of 99.95%. The rate of unknown pattern samples is less than 1/30,000, which can be neglected. The experiment indicates that MSML can also be applied to distributed consistent dataset, due to the fact that MSML does not easily classify known pattern samples into unknown patterns. For the difference between training dataset and test dataset, MSML shows a good adaptability.

3.2 Intrusion Detection Based on Hybrid Multi-Level Data Mining

The development of mobile network is a double-edged sword, which brings both convenience and network security problem to us. Therefore research about network security makes great sense [3]. Traditional network security technologies such as firewall, data encryption, and user authentication system are all passive defense techniques. These methods have a good performance for known attacks, but not for unknown attacks. Different from traditional technologies, Intrusion Detection System (IDS) [15], which is firstly proposed by Denning et al. in 1986 [17], is an innovative and proactive network security technology, meaning that it can detect both known and unknown attacks. The basic architecture of IDS consists of three parts: data acquisition part, intrusion detection part and response process part. Data acquisition part of IDS is used to collect data from internet. The collected data, which consists of normal data and many different types of attacks, is then send to intrusion detection part. Intrusion detection part needs to pick out attack data from normal data and identify what type of the attack is. Response process part receives the detected attacks and processes them according to their types. As the core of IDS, intrusion detection part has become a hot topic in research in recent years.

The widely use of network produces massive data which is a valuable resource. From these data, we can extract much insightful information through data analysis techniques [4, 18, 25, 28]. We name the data collected by data acquisition part 'intrusion data'. Intrusion data has some features that will affect the performance of IDS. We summary these features as follows:

- Multicomponent: The intrusion data is multicomponent because there exists many types of attacks in it. Hence intrusion detection is not a binary-classification problem, but a multi-classification problem, which is more complicated.
- Data imbalance: The number of attacks in different categories vary greatly from each other. It means that the intrusion data has a severe problem of data imbalance. Some attacks are difficult to be detected due to its sparseness.
- Time-varying: The distribution of data is time-varying. The variety between training data and detecting data will affect the performance of IDS.
- Unknown attacks: Some new attacks may appear as time goes by. These unknown attacks are difficult to be detected.

In recent years, researchers begin to use the artificial intelligence (AI) technology to deal with the problems caused by the above mentioned features. Machine learning (ML), whose main idea is building model to mine information from data, is a core subfield of AI including many algorithms [49]. It can be roughly divided into supervised learning algorithms and unsupervised learning algorithms according to whether it has a training phase or not. Supervised learning is the most widely-used technique in machine learning such as *Support Vector Machine* (*SVM*) [19], *Artificial Neural Network* (*ANN*) [20], *Decision Tree* (*DT*) [21], *Random Forest* (*RF*) [22] and so on. Unsupervised learning mainly refers to clustering algorithms [23] such as *K-Means*[24], *DBSCAN*[26], *Affinity Propagation* (*AP*)[27] and so on. Every machine learning algorithm has its own advantages and disadvantages. Combining two or more algorithms together and taking full use of their advantages will increase the performance of IDS. Therefore, how to combine these algorithms scientifically is a noteworthy topic.

In the work of [29], Gang Wang et al. proposed a FCANN framework, which firstly clusters data using *Fuzzy C-means* algorithm, and then classifies each cluster by *ANN*. This framework integrates unsupervised learning and supervised learning to shorten the modelling time, alleviate the data imbalance problem and increase the detection rate. In literature [30], Prasanta Gogoi et al. proposed a MLH-IDS framework which has three layers, including a supervised layer, an unsupervised layer and an outlier-based layer. The supervised layer is used to detect DoS [14] and Probe [14] attacks, and the unsupervised layer is used to detect Normal data. The outlier-based layer is used to distinguish R2L [14] and U2R [14] attacks from each other. The hybrid multi-level framework takes full advantage of different ML algorithms, which is more flexible and has a better performance.

An appropriate scheme of data engineering can also improve the performance of IDS. Data engineering is an indispensable procedure in data mining which includes some widely used techniques such as data preprocessing and dimension reduction [32]. Data preprocessing techniques such as data cleaning and normalization can help remove 'dirty data' and turn data into suitable form. The representative method of dimension reduction is feature selection which is used to remove interfering and redundant features to improve the performance of data mining project. In IDS, the 'intrusion data' is usually not suitable to be detected directly, which needs to be processed by an appropriate data engineering method. Most works we mentioned above mainly focus on the combination of ML algorithms without an elaborately designed data engineering method.

In this part, we propose a novel IDS framework named HMLD through jointly considering data engineering and machine learning. HMLD consists of three modules, including Multi-Level Hybrid Data Engineering (MH-DE) module, Multi-Level Hybrid Machine Learning (MH-ML) module, and Micro Expert Modify (MEM) module. The MH-DE module focuses on data engineering and the MH-ML module focuses on machine learning. These two modules construct a closed cycle of Hybrid Multi-Level Data Mining, which provides a separated and customized detection of different attacks. The hierarchical architecture can address the problems caused by multicomponent and data imbalance. After performing the procedure of MH-DE and MH-ML, most easily detected attacks have been marked. MEM module is used to identify those new attacks which are difficult to detect. The HMLD framework can be implemented in a variety of networks including mobile networks by using different algorithms and parameters.

In this part, we use KDDCUP99 dataset to evaluate the performance of HMLD. Experimental results show that HMLD can achieve 96.70% accuracy which is nearly 1% higher than the recent proposed optimal algorithm *SVM+ELM+Modified K-Means* [13]. Meanwhile, it has a better performance than some other methods in identifying DoS and R2L attacks.

3.2.1 The Framework of HMLD

In this subsection, we will introduce the framework and workflow of HMLD which can detect different categories of attacks separately by different data engineering methods and machine learning methods. The framework of HMLD is illustrated in Fig. 3.11. The input of HMLD contains two parts, one is the labeled training

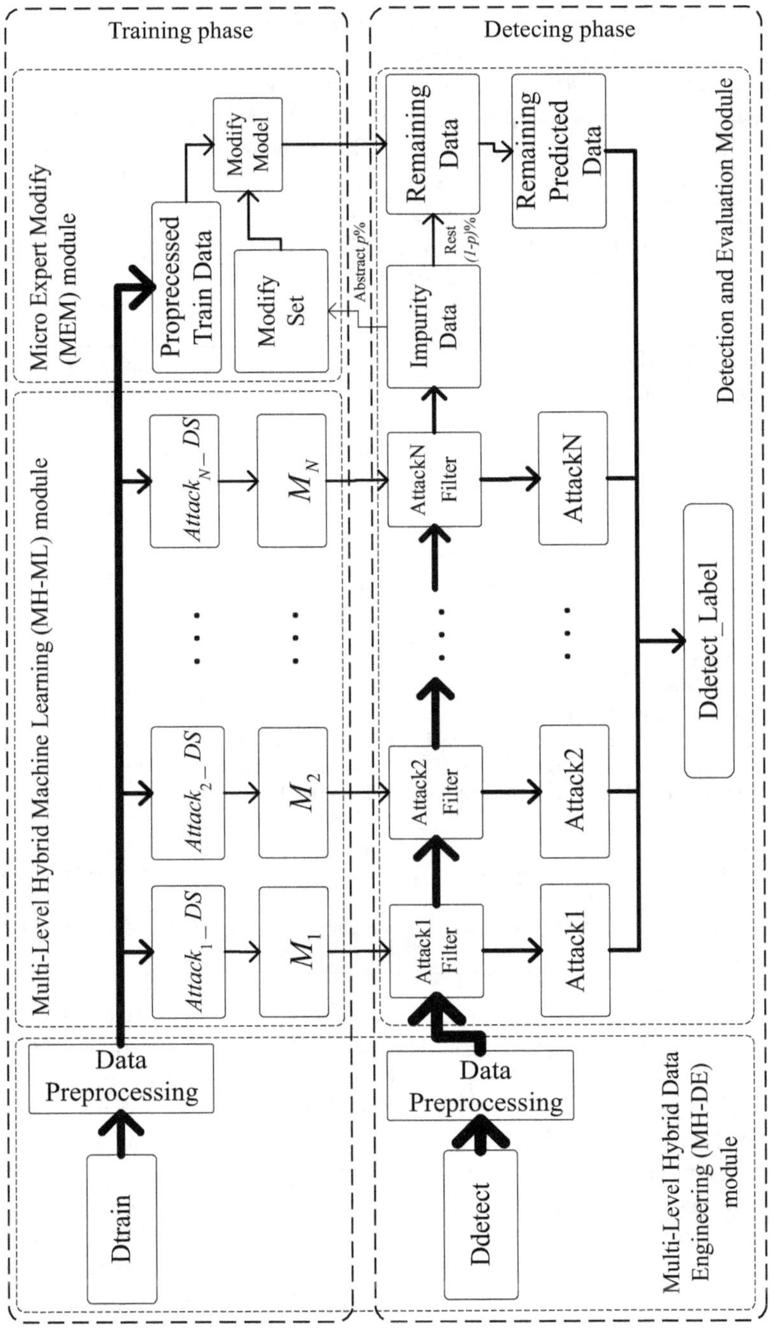

Fig. 3.11 Framework of HMLD

dataset *Dtrain* which is used to train the ML model, the other one is the unlabeled detecting dataset *Ddetect* which is waiting to be detected. *Dtrain* is processed by three modules: MH-DE, MH-ML and MEM as shown in Fig. 3.1, to construct models that are used to detect intrusions in *Ddetect*. The output of HMLD is the labeled detecting dataset *Ddetect_Label*. Algorithm 5 is the pseudo code of HMLD which illustrates the step-by-step workflow of HMLD. We assume that there are N attack categories in input data. The set of attacks in category i, where $i \in [1, N]$, is denoted as S_i. Each attack category has a corresponding package which is denoted as $Attack_i\_DS$. We define the data engineering method used on $Attack_i\_DS$ as D_i, and the ML model trained by $Attack_i\_DS$ is denoted by M_i. $P_{\_key}$, $P_{\_nonkey}$ and P_i are intermediate variables in Algorithm 5.

Algorithm 5 HMLD

1: **Input**: $Dtrain, Ddetect$
2: **Output**: $Ddetect\_Label$
3: Initialization: $D = \{D_i, D_2, \cdots, D_N\}$, $Ddetect\_Label = \emptyset$, $P_0 = Dtrain$
4: **MH-DE module**
5: **for** $i \in [1, N]$ **do**
6: $P_{\_key} = [data.label = i | data \in S_i]$
7: $P_{\_nonkey} = [data.label = 0 | data \notin S_i \& data \in P_{i-1}]$
8: $Attack_i\_DS = do\ D_i\ to\ (P_{\_key} + P_{\_nonkey})$
9: $P_i = P_i - P_{\_key}$
10: **end for**
11: **MH-ML module**
12: **for** $i \in [1, N]$ **do**
13: $M_i =$ the ML model trained using $Attack_i\_DS$
14: Use M_i to detect $Ddetect$ and label the detected attacks as i
15: $Ddetect\_Label = Ddetect\_Label +$ detected attacks with label i
16: $Ddetect = Ddetect -$ detected attacks with label i
17: **end for**
18: **MEM module**
19: Extract $p\%$ data from the remaining data in $Ddetect$ and mark them by experts forming *modify set*
20: Merge *modify set* and $Dtrain$ to retrain a ML model to detect the rest of $Ddetect$ **return** $Ddetect\_Label$

Dtrain is firstly sent to MH-DE module. The 4–10 lines in Algorithm 5 show the workflow of MH-DE module. The line 5 means we will build packages and using data engineering techniques to attacks in category i where $i \in [1, N]$ one by one. The 6–8 lines in Algorithm 5 show how to build package $Attack_i\_DS$. We label the attacks belonging to S_i with i in *Dtrain* and denote this part of data as $P_{\_key}$. Then we label remaining data which is belong to P_{i-1} but not belong to S_i with 0 and denote this part of data as $P_{\_nonkey}$. Thus package $Attack_i\_DS$ is comprised of $P_{\_key}$ and $P_{\_nonkey}$ by using D_i to them which is shown in line 8. This step can make the packages be more suitable for ML modelling by converting format and removing redundancy. Package P_i is constructed by filtering out $P_{\_key}$. We repeat this labeling process to build N packets in sequence.

After MH-DE, these packages are sent to MH-ML module. The 11–17 lines in Algorithm 5 show the workflow of MH-ML module. In MH-ML, each package is used as a training dataset to train an appropriate ML model as shown in line 13 in Algorithm 5. The trained model M_i aims to correctly detect attacks in category i as much as possible. In detecting phase, we use $M_1, M_2, \cdots, M_N$ one by one to mark and filter out attacks in different categories from *Ddetect* which is showed in line 14–16 in Algorithm 5.

After this filtering procedure, the remaining unmarked data in $Ddetect$ are named *impurity data*. *Impurity data* mixes a large amount of normal data with some difficult detected unknown attacks. The 18–20 lines in Algorithm 5 show the workflow of MEM module. We send *impurity data* to MEM module. MEM module samples micro data from the *impurity data* and marks them by experts to form a *modify set*. Then MEM module merges the *modify set* with *Dtrain* to train a new model to identify the unknown attacks in the *impurity data*. After finishing all the procedures, we can get *Ddetect_Label*, which is the result of our detecting work.

3.2.2 *HMLD with KDDCUP99*

The HMLD framework can be applied in many different types of networks by using different algorithms and parameters. In this part, we evaluate the performance of the HMLD framework via KDDCUP99 [14] dataset. This section describes the selection details of algorithms and parameters of each module when using KDDCUP99.

3.2.2.1 KDDCUP99 Dataset

Firstly, we give a brief description of KDDCUP99 dataset. KDDCUP99 Intrusion Detection Dataset is a classic dataset, which has 41 features and 5 types of labels [52], namely Normal, DoS, Probe, U2R, and R2L, respectively. The meaning of the labels are summarized in Table 3.7:

Table 3.7 The definitions of five labels

Label	Meaning
Normal	Normal data
DoS	Denial-of-service
Probe	Surveillance and probing
R2L	Unauthorized access from a remote machine to a local machine
U2R	Unauthorized access to local super-user privileges by a local machine

Table 3.8 The distribution of training data and testing data

Dataset	Normal	DoS	Probe	R2L	U2R	Sum
Training data	9725	39,164	417	111	3	49,420
Testing data	60,593	229,853	4166	16,189	228	311,028

In this part, we extract 10% data from *kddcup.data_10_percent.gz* [14] as training data, and use *corrected.gz* [14] as testing data. Table 3.8 shows the number of each category of attacks.

DoS, Probe, U2R and R2L are four attack categories. Each of them can be further separated into many subcategories. There are 22 subcategories in the training data and 39 subcategories in the testing data. The 17 new subcategories can be considered as new or unknown attack subcategories. Among the four types of attacks, R2L and U2R are much more difficult to be detected than DoS and Probe [30]. R2L is difficult to be detected because it includes many new subcategories of attacks. U2R is difficult to detect due to its sparseness.

3.2.2.2 MH-DE Module

MH-DE module focuses on data engineering which contains the stage of *basic data preprocessing* and the stage of *hybrid feature selection. Basic data preprocessing* is used to make the data suitable for modelling and *hybrid feature selection* is used to remove redundant features.

We design the process of *basic data preprocessing* according to the feature of KDDCUP99 dataset. There are three types of features in KDDCUP99, including factorial type, continuous numerical type and discrete type. Therefore, the *basic data preprocessing* for KDDCUP99 includes the *factorial feature digitizing* procedure and the *continuous feature normalizing* procedure. The former one will map the factorial features into numbers. Without this digital procedure, the dataset cannot be trained by a ML algorithm. The *continuous feature normalizing* procedure normalizes all features to the range of [0, 1]. This step can eliminate the effect caused by the diverse range of features.

In the stage of *hybrid feature selection*, we adopt different feature selection methods for different packages according to the category of attacks. The flow chart of MH-DE is demonstrated in Fig. 3.12. We number the attacks in category DoS, Probe, U2R and R2L as 1, 2, 3, 4, respectively. In MH-DE module, we firstly pick out all the DoS attacks and label them with 1. At the same time, we name the set of other data as 'Other1' and label them as 0. Then do feature selection work to the data. The selected features can distinguish DoS attacks from other data to the utmost. The authors of [34] gave the results of the optimal feature selection subsets of KDDCUP99 dataset, which is depicted in Table 3.9. The numbers in Table 3.9 represent the indexes of features in KDDCUP99. We use DoS feature selection subset in Table 3.9 as the feature selection subset to get DoS_DS. After that, we pick

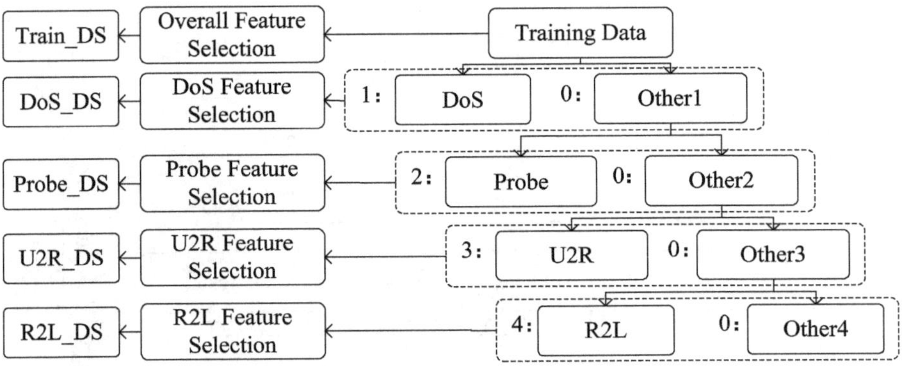

Fig. 3.12 Flow chart of MH-DE

Table 3.9 FMIFS feature selection results of KDDCUP99

Attack	Feature selection subset generated by FMIFS
DoS	2,3,5,6,8,12,23,24,31,32,36,37
Probe	3,4,5,17,19,22,24,25,27,28,29,30,32,33,34,35,37,40,41
U2R	1,2,3,4,6,7,8,12,13,14,15,16,17,18,19,20,21,22,29,31,32,37,40
R2L	1,3,5,6,8,9,10,11,15,17,22,23,24,32,33
Overall	2,3,5,6,9,12,17,23,24,26,29,31,32,33,34,35,36,37,39

out all the Probe attack data from ‘Other1’ and label them with 2. In the meanwhile, we name the remaining data as ‘Other2’ and label them with 0. Then we use Probe feature selection subset in Table 3.9 to get Probe_DS. We repeat the this procedure for U2R and R2L, respectively. After finishing the procedure of MH-DE module, these five packages namely DoS_DS, Probe_DS, U2R_DS, R2L_DS and Train_DS are formed for subsequent training.

3.2.2.3 MH-ML Module

After finishing the MH-DE module, the aforementioned five packages are sent into MH-ML module. Each package needs to build a model to filter out its corresponding category of attacks as many as possible. The authors of [29] proposed a modelling framework including a clustering phase and a classifying phase. In HMLD, we adopt this framework as our modelling framework because it can shorten the modelling time and alleviate data imbalance problem. The basic model building process is shown in Fig. 3.13. In this framework, the training data is firstly clustered by an unsupervised clustering algorithm to get many clusters. Towards each cluster, we train a specific supervised ML classifier.

The selection of algorithms and parameters of MH-ML module needs to be elaborate designed. We will use experiments to choose the algorithms and parameters for MH-ML when using KDDCUP99 dataset.

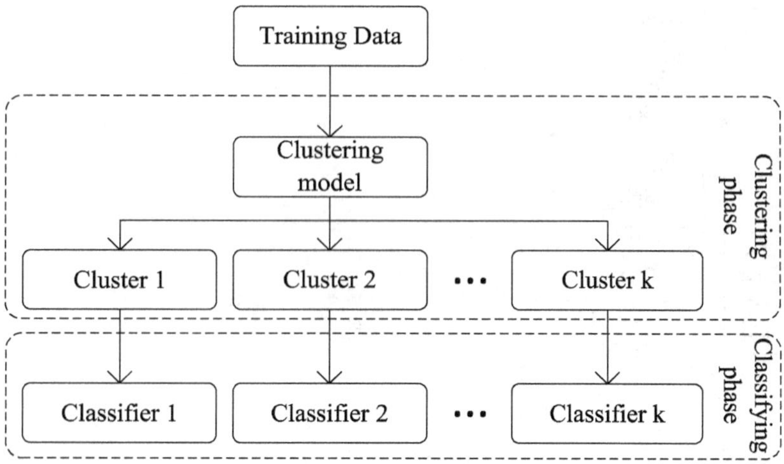

Fig. 3.13 Framework of model building in MH-ML

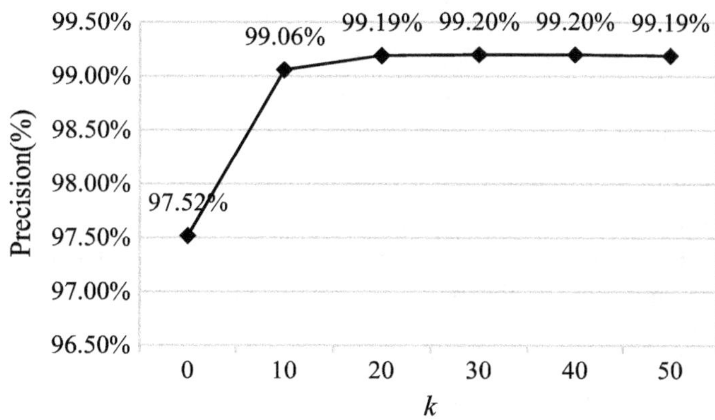

Fig. 3.14 Different k for DoS_DS

In clustering phase, we adopt the *K-Means* [24] algorithm thanks to its good performance and fast computation speed. The main idea of *K-Means* is clustering data into several clusters according to their similarity. We define the number of clusters as k, which has an impact on the performance of HMLD. Figures 3.14, 3.15, 3.16, and 3.17 show the *Precision* of HMLD with k ranging from 0 to 50 for package DoS_DS, Probe_DS, U2R_DS, R2L_DS, respectively. The *Precision* of detecting attacks represents the proportion of predicted attacks which are actually attacks. When k is 0, it means that we do not use the *K-Means*. DoS attacks can reach highest *Precision* when k is equal to 30 and 40. Therefore, we set k to 30 for DoS attacks because a smaller k can reduce computing resources and shorten modelling time. Probe attacks can achieve 91.91% *Precision* when k is 0 and 92.08% when k is 20. Though the *Precision* is a bit lower when k is 0, the modelling complexity can

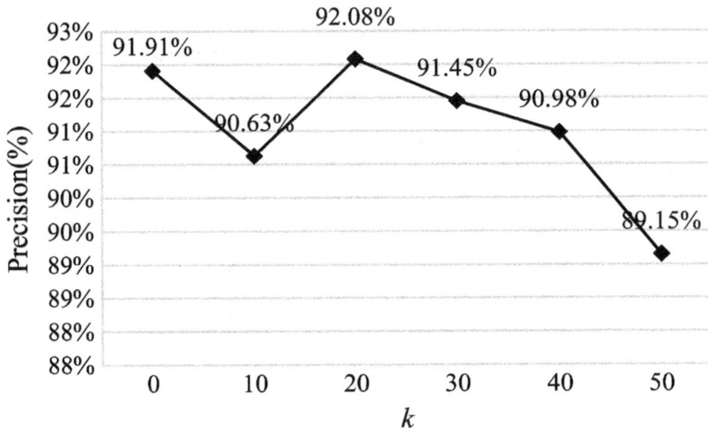

Fig. 3.15 Different k for Probe_DS

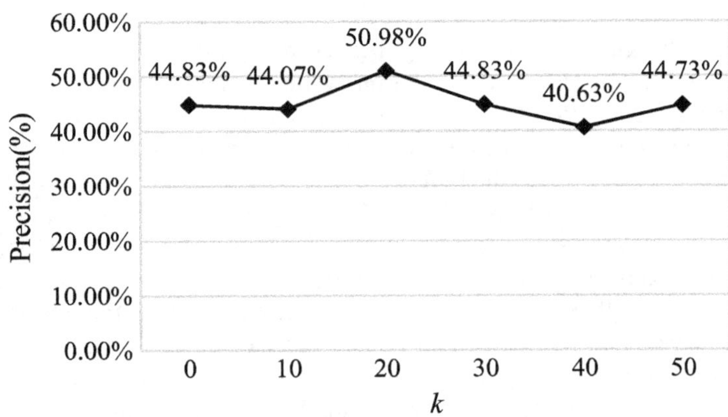

Fig. 3.16 Different k for U2R_DS

be significantly reduced. Hence we set k to 0 for Probe attacks. We set k to 20 and 30 for U2R and R2L attacks respectively to maximize the performance, which can be observed from Figs. 3.16 and 3.17.

In classifying phase, we compare many ML algorithms including *SVM*[19], *ANN* [20], *DT*[21] and *RF*[22] with different parameters. These algorithms are all supervised learning algorithms which are mainly used in classification problems. The main idea of *linear SVM* is to find a support hyperplane which can separate different types of data in best performance. If the data can be separated much better using a hypersurface than a hyperplane, we can use *kernel SVM* to replace *linear SVM*. The most frequently-used kernel of *SVM* is *rbf* kernel, which has another name as *Gaussian* kernel. Parameter r is the variance of the *rbf* kernel. Parameter C is the relaxation factor of *SVM* [56]. Smaller C will cause a decline of detection accuracy but can alleviate the overfitting problem. *ANN* is a multilayer

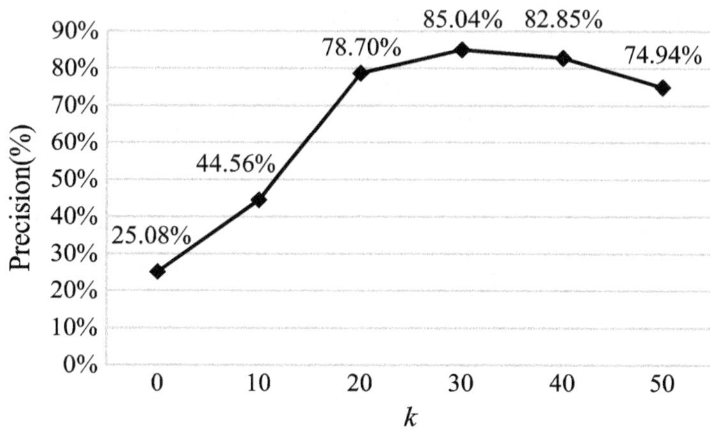

Fig. 3.17 Different *k* for R2L_DS

neural network which contains an input layer, many hidden layers and an output layer. Each layer consists of many neurons which contains many parameters such as weights, bias and activation functions. The activation functions can be *identity*, *logistic*, *tanh* or *relu*. *DT* algorithm computes the *information gain* of each feature and selects the biggest one as the root of a tree. Then repeat this procedure iteratively until the stop condition is satisfied. *CART* is a representative algorithm of *DT*. *RF* is a integration of *DT*, which samples data and features many times to build many trees. Then *RF* gets the final results by comprehensively considering all these trees.

Figures 3.18, 3.19, 3.20, and 3.21 show the number of correctly detected attacks by using different algorithms with different parameters. From these figures, we can see that the performance of different algorithms varies greatly from each other for different attacks. We choose the appropriate algorithm according to two metrics, the number of detected attacks and the *Precision* of detecting attacks. The number of detected attacks can be observed from Figs. 3.18, 3.19, 3.20, and 3.21. A better algorithm can detect more attacks and can get a larger *Precision*.

For DoS attacks, *SVM-linear (C = 1.0)*, *ANN-identity*, *ANN-logistic* and *ANN-tanh* all can detect more attacks than other algorithms. The *Precision* of DoS attack is 99.20% when using *SVM-linear* which is the highest among these four algorithms. Hence, we select *SVM-linear (C = 1.0)* as the classification algorithm of DoS_DS. For Probe attacks, *ANN-logistic*, *CART* can detect more attacks than others. Comparing their *Precision*, *CART* is 69.40% and *ANN -logistic* is 92.62%. For Probe attack, *ANN-logistic* algorithm is better. For U2R attacks, *ANN-tanh* and *ANN-identity* can detect more attacks than others. When using *ANN-tanh*, the *Precision* of U2R is 7.28%. But when using *ANN-identity*, the *Precision* is increased to 37.14%. Thus, we choose *ANN-identity* for U2R_DS. For R2L attacks, experiments show that *SVM-rbf* and *ANN-relu* have a better performance. However, the *Precision* is 84.2% when using *ANN-relu*, and 82.26% when using *SVM-rbf*. Therefore, we use *ANN-relu* algorithm for modelling of R2L_DS. The design of MH-ML module is shown in Fig. 3.22.

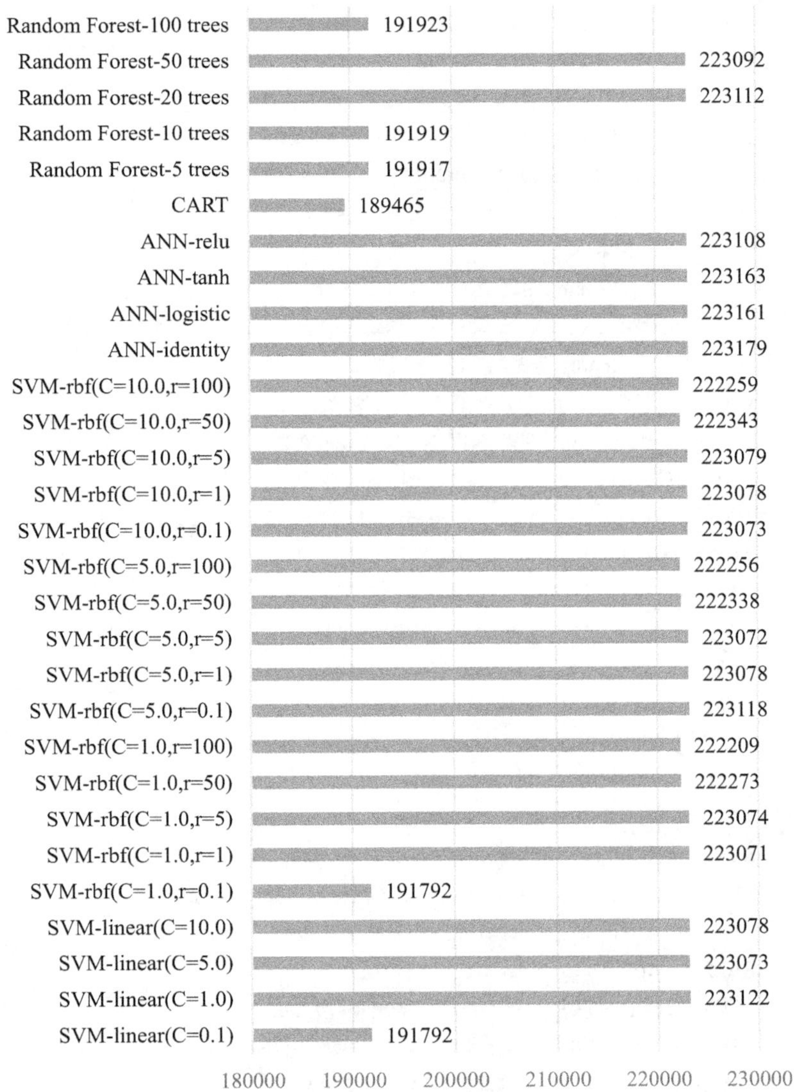

Fig. 3.18 Correctly detected DoS attacks

3.2.2.4 MEM Module

After MH-ML, most of the attacks are marked and filtered out. We name the remaining unmarked data *impurity data*. The *impurity data* mixes a large amount of normal data with some unknown attacks. Before MEM, if we mark all the *impurity data* as Normal, we can derive the 'confusion matrix' as Table 3.10. In a 'confusion

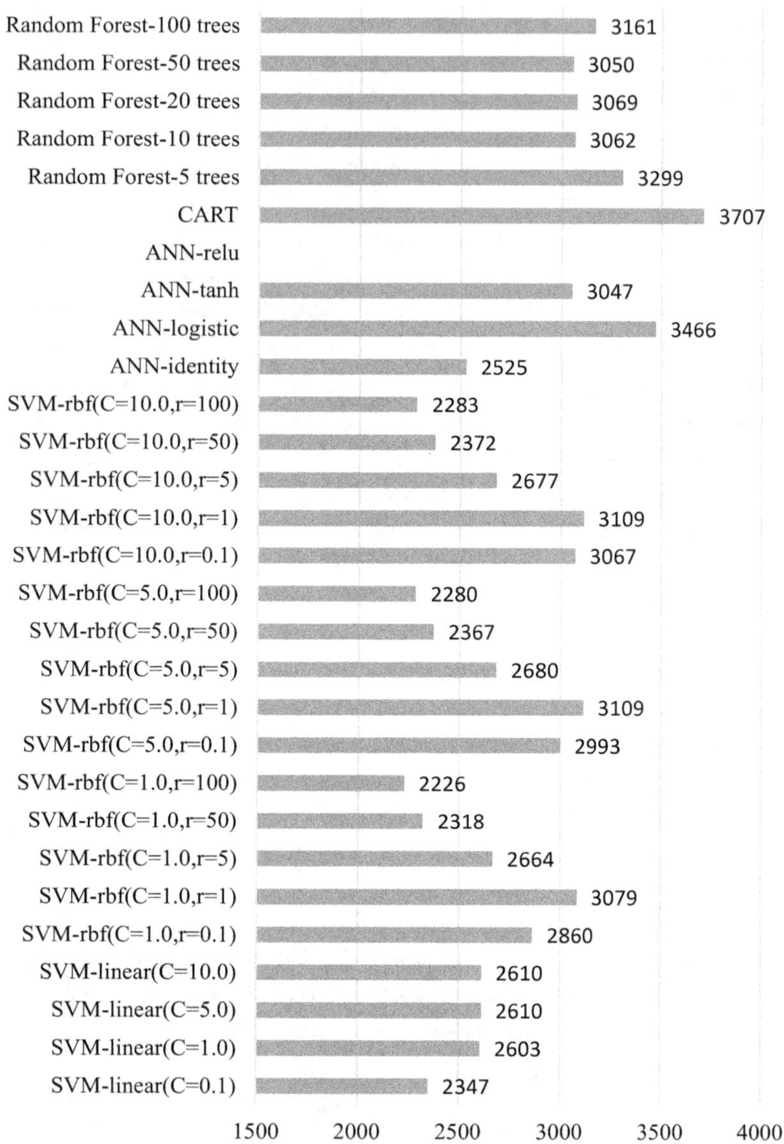

Fig. 3.19 Correctly detected probe attacks

matrix', each row represents the number of data which is actually this type, and each column represents data which is predicted as this type. For example, the number in the upper left corner is the number of data which is actual normal and also be predicted as normal. We can observe from Table 3.10 that a large amount of DoS attacks and R2L attacks are wrongly detected before MEM module because there

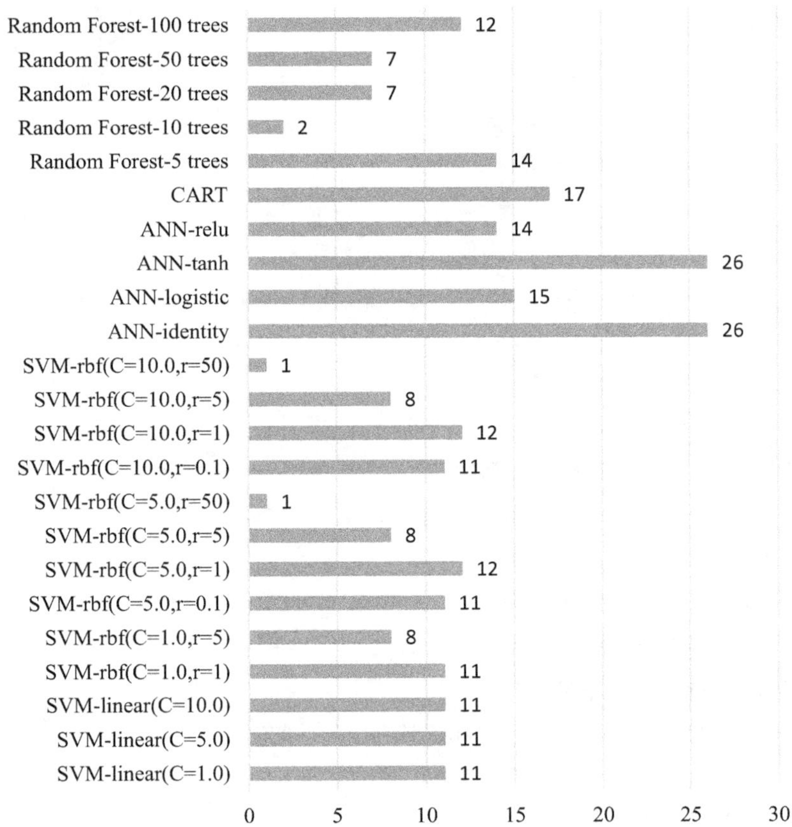

Fig. 3.20 Correctly detected U2R attacks

appears many new subcategories of DoS and R2L attacks. Given that we have no information about these new attacks in training data, these new attacks are difficult to detect.

In order to detect these new attacks efficiently, we send *impurity data* to MEM module. MEM module randomly samples $p\%$ of *impurity data* and marks them to construct the *modify set*. We use *DT* [21] to retrain a ML model thanks to its rapid modelling speed. This model is used to detect the new attacks in *impurity data*. We experiment different value of p to compare the average accuracy of HMLD, and the results are shown in Table 3.11. When $p\%$ is 0.3%, HMLD achieves a relative high performance. When $p\%$ is larger than 0.3%, the growth starts to slow down. Therefore, we set p to 0.3.

MEM module samples 0.3% data from *impurity data* and marks them by experts to form the *Modify Set*. Experimental results show that there are about 240 records in the *Modify Set*. After MEM, the 'confusion matrix' is showed in Table 3.12. With this micro cost, most of the unknown attacks of DoS and R2L are correctly detected.

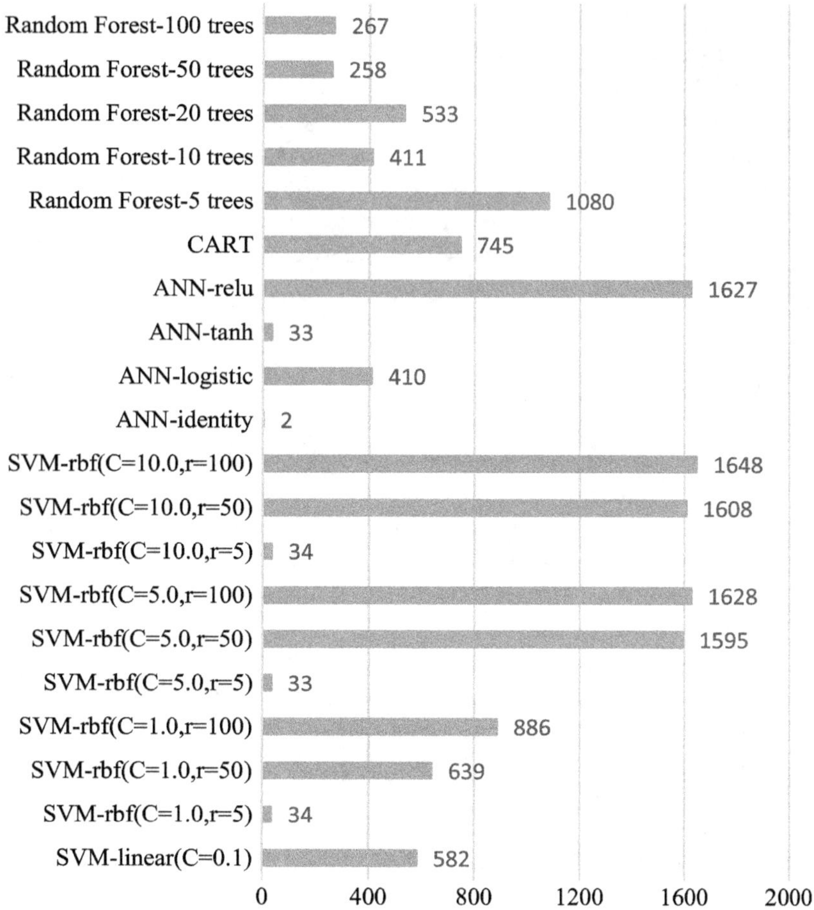

Fig. 3.21 Correctly detected R2L attacks

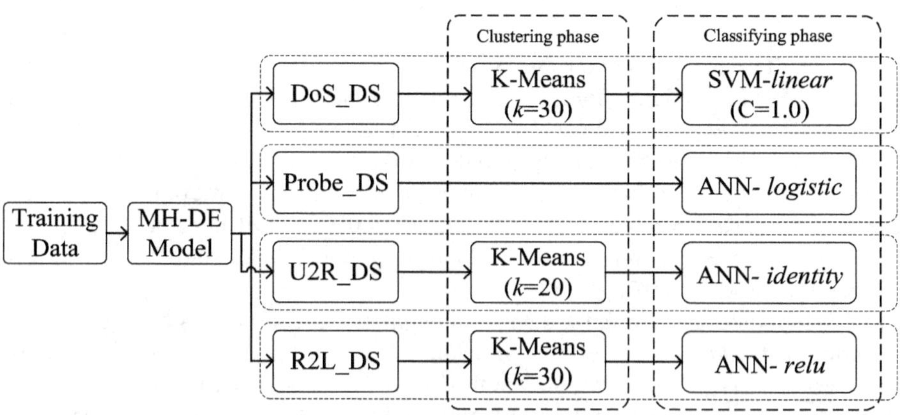

Fig. 3.22 Parameters and algorithms selection for MH-ML

Table 3.10 Confusion matrix before MEM

Actual vs predict	Normal	DoS	Probe	R2L	U2R
Normal	59,391	711	202	272	17
DoS	6674	2,231,222	57	0	0
Probe	358	341	3466	0	1
R2L	13,911	617	28	1626	7
U2R	48	122	18	14	26

Table 3.11 Comparisons of different *p*

p	0.01%	0.05%	0.1%	0.2%	0.3%	0.4%	0.5%
Accuracy	92.56%	95.12%	96.00%	96.21%	96.50%	96.60%	96.62%

Table 3.12 Confusion matrix after MEM

Actual vs predict	Normal	DoS	Probe	R2L	U2R
Normal	56,210	763	697	2699	39
DoS	192	229,565	70	6	0
Probe	34	481	3615	35	1
R2L	4333	620	86	11,102	9
U2R	28	122	32	20	26

3.2.3 *Experimental Results and Discussions*

3.2.3.1 Evaluation Criteria

TP(true positives), TN(true negatives), FP(false positives), FN(false negatives) [34] are usually used to evaluate the performance of IDS. In IDS, we define normal data as positive and define attacks as negative. TP is the number of data which is actual positive and also predicted as positive. TN is the number of data which is actual positive but predicted as negative. FP is the number of data which is actual negative but predicted as positive. FN is the number of data which is actual negative and also predicted as negative. TP, TN, FP and FN can be written in a 'confusion matrix' as shown in Table 3.13.

Precision, Recall, F-value and *Accuracy* [34] are also defined to evaluate the performance. *Precision* given by Eq. (3.11) is the proportion of predicted positives values which are actually positive. *Recall* given by Eq. (3.12) is the proportion of the actual number of positives which are correctly identified. *Recall* also can be called as *Detection rate*. *F-value* given by Eq. (3.13) is a harmonic mean between

Table 3.13 Confusion matrix

	Predict as positive	Predict as negative
Actual is positive	TP	TN
Actual is negative	FP	FN

Precision and *Recall*. *Accuracy* given by Eq. (3.14) is the proportion of correctly predicted data.

$$P = Precision = \frac{TP}{TP + FP}, \tag{3.11}$$

$$R = Recall = Detection\ rate = \frac{TP}{TP + FN}, \tag{3.12}$$

$$F - value = \frac{2 \times Precision \times Recall}{Precision + Recall}, \tag{3.13}$$

$$Accurary = \frac{TP + TN}{TP + TN + FN + FP}. \tag{3.14}$$

3.2.3.2 Experiments and Analysis

(1) Performance of HMLD

In this part, we show the overall performance of HMLD in terms of evaluation indicators including *Precision, Recall, F-value* and *Accuracy*. Table 3.14 shows the *Precision*, *Recall* and *F-valve* of each attack. The *Accuracy* is 96.70% which can be computed by Eq. (3.14).

(2) Comparisons of *Hybrid Feature Selection* Used by ML-DE with Other Feature Selection Methods

In the ML-DE module of HMLD, we adopt *hybrid feature selection* as the feature selection method in data engineering. In contrast, we tested another two commonly used feature selection methods. One is doing the same feature selection to all packages, which called *unified feature selection*. When do *unified feature selection*, we use the feature subset of *Overall* in Table 3.9 to all packages. The other is not to do feature selection, which called *no feature selection*. Table 3.15 and Fig. 3.23 shows the comparisons of TP and *Precision* of different attacks by using *hybrid feature selection*, *unified feature selection* and *no feature selection*.

We can see that the overall performance of *hybrid feature selection* is much better than the other methods. From Table 3.15, the numbers of correctly detected attacks using *hybrid feature selection* are the highest for DoS, Probe and R2L. For U2R, the number is just 1 less than *no feature selection*. In terms of *Precision*, from Fig. 3.23

Table 3.14 Indicators of HMLD

Indicators (%)	Normal	DoS	Probe	R2L	U2R	Overall
Precision	92.46	99.14	80.33	80.10	34.67	96.55
Recall	93.05	99.88	86.77	68.74	11.40	96.70
F-value	92.75	99.51	83.42	73.98	17.16	96.60

Table 3.15 Numbers of correctly detected attacks

No. of correctly detected attacks	DoS	Probe	U2R	R2L
Hybrid feature selection	**223,122**	**3466**	26	**1626**
Unified feature selection	222,630	2662	4	748
No feature selection	208,424	3355	**27**	252

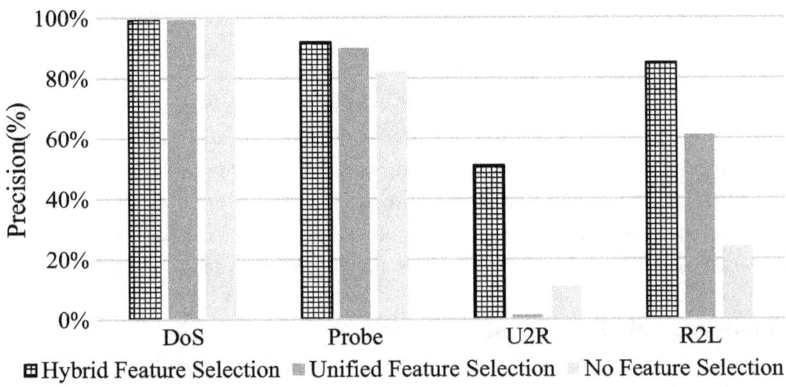

Fig. 3.23 Comparison of *Precision* for different attacks by using different feature selection methods

we can see that, the *Precision* using *hybrid feature selection* is much higher than the other two methods for Probe, R2L and U2R. For DoS, the *Precision* is almost in the same level. The reason is that the characteristic of different attacks is reflected in different subset of features. Using customized subset of features can improve the performance on the corresponding attacks. Therefore, we can conclude that *hybrid feature selection* is far superior to other methods.

(3) Comparisons of *Hybrid Multi-Level ML* Used by MH-ML with Single ML Methods

In MH-ML of HMLD, we use the method of selecting different ML algorithms for different packages according to their corresponding category of attacks. This method is called *Hybrid Multi-Level Machine Learning* method. In the meanwhile, we call the method of using the same ML algorithm such as *SVM*, *ANN* and *RF* to all packages as *Single Machine Learning* method. Figure 3.24 contrasts the performance of using *Hybrid Multi-Level Machine Learning* and *Single Machine Learning*. From Fig. 3.24 we can observe that the *Recall* and *F-value* of using *Hybrid Multi-Level Machine Learning* method is much better than using *Single*

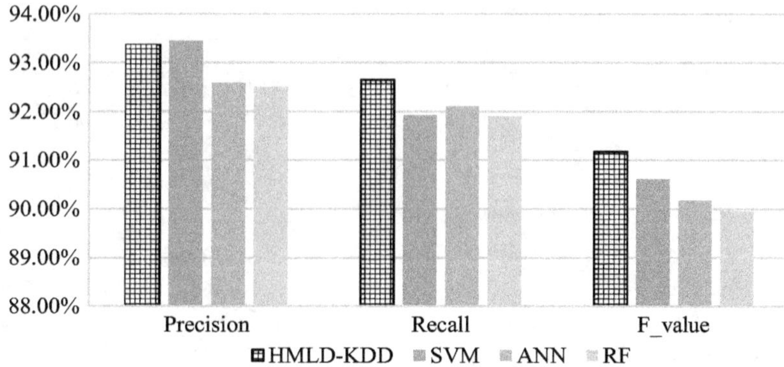

Fig. 3.24 Performance comparison of *Hybrid multi-level machine learning* and *Single machine learning*

Machine Learning methods. The *Precision* of using *Hybrid Multi-Level Machine Learning* method is higher than using *ANN* and *RF*, but a bit less than using *SVM*. Different algorithms has a different performance on detecting different attacks and *Hybrid Multi-Level Machine Learning* choose suitable algorithms to each category of attacks. Hence, the overall performance of using *Hybrid Multi-Level Machine Learning* is much better than using *Single Machine Learning*.

(4) Comparisons of HMLD with Prior Works

In this subsection, we compare the performance of HMLD with some prior works [13, 35, 37] proposed in recent years. The authors of literature [13] proposed a modified *K-Means* to build a high-quality training dataset and train a multi-level hybrid intrusion detection model using *SVM* and *ELM*. The authors of literature [35] proposed an IDS based on *SVM*, hierarchical clustering algorithm and a simple feature selection procedure. The literature [37] is the method proposed by the winner of KDDCUP99 competition.

Tables 3.16 and 3.17 show comparisons between HMLD and these aforementioned works. In Table 3.16, each row represents the detection rates of different data types generated by one algorithm. Each column represents the detection rates of one data type by using different algorithms.

According to Table 3.16, HMLD has a high *Detection rate* of DoS and R2L attacks. Though the *Detection rate* of Normal and Probe is not the highest, they have the rates at or just below the average. The *Detection rate* of U2R is low, which is because the number of U2R attacks in training data is too small. From Table 3.12, we can see that most of the U2R attacks are predicted as DoS attacks. Though detected as wrong category, it is still classified as an attack category, not a normal class. From Table 3.17, we can observe that the overall performance of HMLD is better than the prior works, with a detection *Accuracy* reaching 96.70%, which is nearly 1% higher than the optimal in the other. In summary, the overall performance of HMLD model is better than prior works.

Table 3.16 *Detection rate (%)* generated by HMLD and some prior works

Algorithms	Normal	DoS	Probe	R2L	U2R
HMLD-KDD	93.05	**99.88**	86.77	**68.74**	11.40
SVM+ELM+Modify K-Means	98.13	99.54	87.22	21.93	**31.79**
SVM+BIRCH clustering	99.30	99.50	**97.50**	19.70	28.80
Winner of the KDDCUP99	**99.50**	97.10	83.30	13.20	8.40

Table 3.17 *Accuracy (%)* comparisons of HMLD and some prior works

Algorithms	Accuracy(%)
HMLD-KDD	**96.70**
SVM+ELM+Modify K-means	95.75
SVM+BIRCH clustering	95.70
Winner of the KDDCUP99	93.30

3.3 Abnormal Network Traffic Detection Based on Big Data Analysis

With the improvement of network, storage, calculation and transmission,the Internet is interacting more closely with people than ever before. While the Internet is making our life more convenient, it brings about some potential risks [6]. For example, malicious attacks involving user privacy and security become more and more frequent.

The changes of how people use the Internet is a new challenge for traditional abnormal network event detection techniques. It is more hard for researchers to get aware of some new kinds of attacks [62]. To resolve these problems, some abnormal network traffic detection methods have been proposed. Traditional abnormal traffic detection method can be classified into two categories [38, 41, 42, 47]. One is misuse detection, and the other is abnormal detection. The two methods have their own pros and cons. Misuse detection has a high accuracy but needs support from known knowledge. Abnormal detection do not need known knowledge, but cannot categorize the type of attacks, the accuracy is also lower. For example, Hari Om [44] designes a hybrid detection system, which is a hybrid anomaly detection system considering k-means, k-nearest neighbor and Naïve Bayes methods.

However, the explosive increase of network traffic has directly or indirectly pushed the Internet into the big data era, which makes anomaly traffic detection more difficult to deal with because of high calculation volume and constant changes of network data distribution caused by big data [45, 46, 48, 50]. Because the speed of network data generation is fast, it makes the volume of normal traffic and abnormal traffic differ a lot, and the distribution of the data change. Besides, with big data, the difference between normal traffic and abnormal traffic is increasing [55, 58]. It makes the traditional methods unable to effectively detect abnormal traffic.

Therefore, to increase the accuracy of abnormal traffic and avoid the loose caused by false negative detection, we propose a novel model based on big data analytics, which can avoid the influence brought by adjustment of network traffic distribution, increase dectection accuracy and reduce the false negative rate. The core of the proposed model is not simply combination of traditional detection methods, but a novel detection model based on big data. In the simulation, we use k-means, decision tree and random forest algorithms as comparative objects to vertify the effectiveness of our model. Simulation results reveal that the proposed model has a much better performance, which can achieve a detection rate of 95.4% on normal data, 98.6% on DoS attack, 93.9% on Probe attack, 56.1% on U2R attack, and 77.2% on R2L attack.

3.3.1 System Model

Influenced by big data, network data distribution is gradually changing [5]. This part try to solve the problem that caused by the increasing difference between normal traffic and abnormal traffic. Therefore, we proposed a new abnormal traffic detection model based on big data analysis, and this model includes three sub-models.

3.3.1.1 Normal Traffic Selection Model

Normal traffic selection model uses classification and clustering algorithm to distinguish normal and anomaly behaviors, rather than involved specific anomaly behaviors. This model includes two stages:

1. Training stage: training model uses data that labeled normal or abnormal, and the model applies in test stage.
2. Test stage: test stage is similar to detection in practice. Using unlabeled date, the model classifies traffic data into normal or abnormal, and labels them.

Normal traffic selection model uses k-means clustering algorithm, KNN, decision tree and random forest classification algorithms. Traditionally, before using k-means algorithms, it is very important to set the number of categories, because we don't know how many categories. But in order to distinguish normal and abnormal behavior, the normal traffic selection model uses k-means as following way.

In training stage, using labeled information classify data into normal and abnormal. These two categories use k-means separately instead of clustering all data at once, getting the center of the data set respectively. Then using the center of the data set, KNN clustering algorithm classifies test data. Decision tree and random forest classification algorithms train with labeled normal and abnormal data.

3.3.1.2 Abnormal Traffic Selection Model

The purpose of abnormal traffic selection model is avoid influence caused by too many normal traffic than abnormal traffic. This model classifies anomaly traffic into specific categories, and includes two stage as well:

1. Training stage: this stage only use abnormal data to train classification model, and every data label specific attack group. Using classification algorithms learns classified rules.
2. Test stage: test stage is similar to detection in practice, using unlabeled data (including normal behavior data). The classification model classifies anomaly traffic into specific categories according to the classified rules, and gives specific label to every data.

Abnormal traffic selection model uses decision tree and random forest classification algorithms. Abnormal traffic selection model and normal traffic selection model are independent, without order of priority in training stage or test stage.

3.3.1.3 Abnormal Traffic Selection Model

Mixed compensation model combines the result from normal traffic selection model and abnormal traffic selection model to produce a final result. Although abnormal traffic selection model is more effective because without influence of normal traffic data, the model has high false negative rate due to this characteristic. Therefore use normal set produced by normal traffic selection model to compensate abnormal set $A = \{A_1, A_2, \cdots, A_k\}$ produced by abnormal traffic selection model. $A_i, i \in [1, k]$ denote specific attack category. If c denote detection result, rule of compensation as follow:

$$\begin{cases} \text{if } c \in A_i, c \in N, \ \text{then } c \in N \\ \text{if } c \in A_i, c \notin N, \ \text{then } c \in A_i. \end{cases} \tag{3.15}$$

3.3.2 *Simulation Results and Discussions*

Before using three sub-models of anomaly detection based on big data analysis, data set needs be preprocessed with label for training model. It should be noted that rightly selecting feature is a good way to reduce dimension and increase efficiency of running. In the simulation, three different algorithms are used to verify validity of the proposed model.

3.3.2.1 Data Set

In the simulation, we use KDDCUP99 [53] data set to test my model. KDDCUP99 data set is widespread use for testing abnormal detection model, which is obtained and processed from KDDCUP99 [54]. KDDCUP99 data set has 41 features and been sorted into three group: basic feature, content feature and time feature [57].

The distribution of data set is shown as Table 3.18, where training data has 5 million records, 10% of training data has 494,021 records, and test data has 311,029 records. Every record is labeled to be normal or abnormal, and abnormal data can be classified into four groups: Dos, U2R, R2L and Probe. From Table 3.18, we find that normal data in training data set is more than abnormal data in test data set. Therefore, this data set can be used to test the performance of the proposed model under different circumstances.

3.3.2.2 Simulation Results

As shown in Table 3.19, we have done eight experiments with the model based on big data analysis, and three control experiments which used k-means, decision tree or random forest respectively. In the control groups, training classify model uses all training data set with five categories, then classifying test data into five categories. Another control group is winner of KDDCUP99.

Table 3.18 Distribution of KDDCUP99 data set

Data set	Normal	DoS	Probe	R2L	U2R
10% of training data set	97,278	391,458	4107	1126	52
Test data set	60,593	229,853	4166	16,189	228

Table 3.19 Number of experiments

No.	Normal traffic selection model	Abnormal traffic selection model	No. of control group	Algorithm
1	k-means1[a]	Random forest	9	k-means
2	k-means1[a]	Decision tree	10	Decision tree
3	k-means2[a]	Random forest	11	Random forest
4	k-means2[a]	Decision tree	12	Winner of KDDCUP99
5	Decision tree	Decision tree		
6	Decision tree	Random forest		
7	Random forest	Decision tree		
8	Random forest	Random forest		

[a]In the normal traffic selection model, the number of cluster of normal and abnormal respectively is 4 and 30 in $k\text{-}means1$, and the number of cluster of normal and abnormal respectively is 100 and 300 in $k\text{-}means2$

Table 3.20 Prediction accuracy

No.	Experiment	Normal	DoS	Probe	U2R	R2L
1	k-means1+Random forest	0.632	0.814	0.939	0.561	0.679
2	k-means1+Decision tree	0.656	0.791	0.878	0.500	0.772
3	k-means2+Random forest	0.945	0.983	0.910	0.513	0.510
4	k-means2+Decision tree	0.946	0.979	0.852	0.500	0.504
5	Decision tree+Decision tree	0.951	0.984	0.829	0.500	0.512
6	Decision tree + Random forest	0.951	0.986	0.831	0.550	0.517
7	Random forest + Decision tree	0.954	0.980	0.861	0.500	0.521
8	Random forest + Random forest	0.952	0.985	0.872	0.520	0.510
9	k-means	0.938	0.968	0.785	0.500	0.528
10	Decision tree	0.951	0.983	0.793	0.500	0.500
11	Random forest	0.952	0.985	0.875	0.522	0.507
12	Winner of KDDCUP99	0.995	0.971	0.833	0.132	0.084

Table 3.21 Compared with winner of KDDCUP99

No.	Experiment	Normal	DoS	Probe	U2R	R2L	Final score	Rank
8	Random forest+Random forest	3	2	2	2	4	13	1
6	Decision tree+Random forest	4	1	6	1	2	14	2
7	Random forest +Decision tree	2	5	3	4	1	15	3
2	k-means2+Random forest	7	4	1	3	5	20	4
5	Decision tree+Decision tree	4	3	7	6	3	23	5
4	k-means2+Decision tree	6	6	4	5	6	27	6
12	Winner of KDDCUP99	1	7	5	7	7	27	6

In the simulation, prediction accuracy is used as a simulation metric of detection effect, which is shown in Table 3.20. Besides, we adopt way of sorting and grading for every type. For example, all experiments are sorted by prediction accuracy of normal. The first grades 1 point, the second grades 2 points, and so on. Finally, adding grade of five groups is final grade.

As shown in Table 3.21, the experiment group and winner of KDDCUP99 are sorted by final grade. While the later has high detection rate in normal data, as for four attack types, the result of model based on big data analysis is better than winner of KDDCUP99.

Algorithm of winner of KDDCUP99 is C5 decision tree [59–61, 63]. Training data of winner of KDDCUP99 is a little different with my experiment. Thus for evaluating detection effect of the proposed mode, we did three control experiments with same training data and test data, used with k-means, decision tree or random forest respectively. The number of these experiments is noted as 11, 10 and 9.

Sorting result shows that detection effect of algorithm of the proposed model is better than no use, as shown as Table 3.22. We will discuss experiments results, compared No. 8 with No. 11, No. 7 with No. 5 and No. 3 with No. 4.

Table 3.22 Compared with control group

No.	Experiment	Normal	DoS	Probe	U2R	R2L	Final score	Rank
6	Decision tree+Random forest	4	1	6	1	3	15	1
8	Random forest + Random forest	2	2	3	3	5	15	1
11	Random forest	2	2	2	2	7	15	1
7	Random forest + Decision tree	1	7	4	5	2	19	4
3	k-means2+ Random forest	8	5	1	4	5	23	5
5	Decision tree + Decision tree	4	4	7	5	4	24	6
10	Decision tree	4	5	8	5	9	31	7
9	k-means	9	9	9	5	1	33	8
4	k-means2+ Decision tree	7	8	5	5	8	33	8

3.3.2.3 Discussing Result of No. 8 and No. 11

Score of top three are same. Judging No. 8 and No. 11 with final grade, detection result of two experiments are almost same. And both of them use random forest algorithm. But the difference is:

1. Importance of variable used in classifying is different;
2. No. 8 has lower false negative rate.

- **Importance of Variable**

As shown as Fig. 3.25, variables chosen by random forest in No. 8 and No. 11 are different. Random forest algorithm can output importance of variables, noted Gini index [51]. Figure 3.25 shows that top 20 have important variables in comparison with top 1, whose value is higher and more important.

In No. 8, rank of variables is different between normal traffic selection model and abnormal traffic selection model. This means that variable used for predicting normal or abnormal and specific attack is different. Therefore, choosing variable in No. 11 is influenced by both sides, and output a compromised result when choosing variables, that's why prediction of model in No. 11 has deviation.

- **Comparison of False Negative Rate**

In order to evaluate effect on predicting abnormal behavior, false negative rate is used as an important index, which can measure how many attack events are omitted. Table 3.23 shows confusion matrix of results of experiments No. 8 and No. 11 when using random forest. Row express information of prediction, and column express actual information. False negative rate of No. 8 in normal type is very low, but high in U2R and R2L type. In No. 8, false negative rate of normal selection model in normal is low. Without influence of normal training data, false negative rate of abnormal selection model in four specific attack types are lower than No. 11.

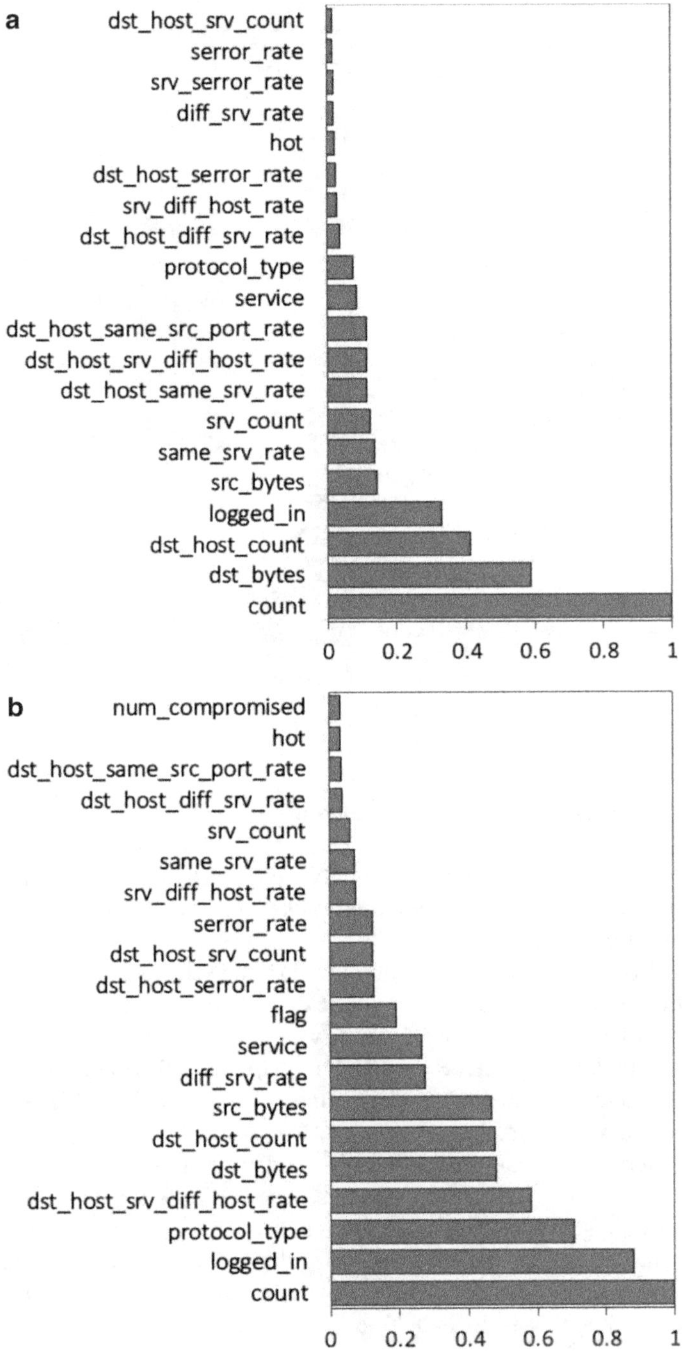

Fig. 3.25 Importance of variables in random forest. (**a**) No. 11. (**b**) No. 8 normal traffic selection model. (**c**) No. 8 abnormal traffic selection model

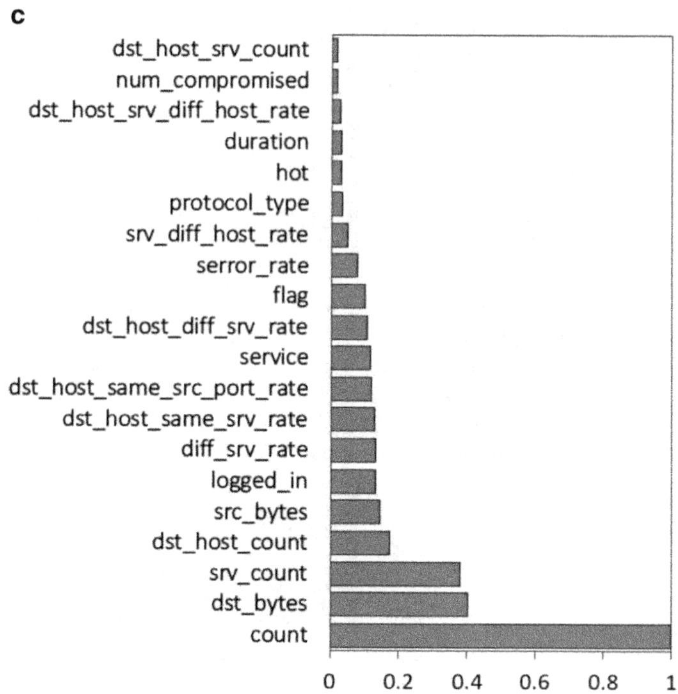

Fig. 3.25 (continued)

3.3.2.4 Discussing Result of No. 5 and No. 7

No. 5 and No. 7 respectively compare with No. 6 and No. 8 by using same algorithm in normal traffic selection model, and their ranks are lower when using decision tree in abnormal traffic selection model.

Table 3.24 is confusion matrix of abnormal traffic selection model with decision tree algorithm. It shows that U2R can not be detected and false negative rate of R2L is higher. In order to find the reason, classify tree is checked in Fig. 3.26, where the classification model prefers DoS and Probe attack, then R2L attack, and no result point of U2R attack. Distribution of training data can explain this phenomenon, which can be shown in Fig. 3.27.

When generating decision tree, the obtained information will cause results in favor of feature which have more samples. Therefore, if the number of training data set in every group is different enough, it cannot get efficient classification model for small samples. Moreover, because the number of between training data is comparatively equal, classification result is better, such as No. 6, when normal traffic selection model uses decision tree.

Table 3.23 Confusion matrix

No. 11					
Prediction	*Normal*	*DoS*	*Probe*	*U2R*	*R2L*
Normal	60,287	5967	847	159	15,839
DoS	69	223,814	191	8	0
Probe	233	72	3128	50	104
U2R	1	0	0	10	5
R2L	3	0	0	1	241
False negative	0.00505	0.026273	0.24916	0.95614	0.985113

No. 8 normal traffic selection model		
Prediction	*Normal*	*Abnormal*
Normal	60,289	22,853
Abnormal	304	227,583
False negative	0.005017	0.091253

No. 8 abnormal traffic selection model				
Prediction	*DoS*	*Probe*	*U2R*	*R2L*
DoS	229,231	769	20	4693
Probe	297	3393	135	5646
U2R	0	0	39	32
R2L	325	4	34	5818
False negative	0.002706	0.18555	0.828947	0.64062

Table 3.24 Confusion matrix of abnormal traffic selection model with decision tree

Predition	DoS	Probe	U2R	R2L
DoS	227,792	589	34	6245
Probe	1434	3192	20	283
U2R	0	0	0	0
R2L	627	385	174	9661

3.3.2.5 Discussing Result of No. 3 and No. 4

No. 3 and No. 4 use k-means in normal traffic selection model to choose clustering center. Table 3.25 shows final prediction accuracies in No. 3 and No. 4. Because final results are lower than that of normal traffic selection model or abnormal traffic selection model, we find that this problem is caused by using k-means in normal selection model. Table 3.26 shows confusion matrix of normal traffic selection model of No. 3 and No. 4. Many abnormal records are predicted as normal, which cause high false negative rate. Therefore, many abnormal records predicted by abnormal traffic selection model will be regarded as normal after mixed compensation model.

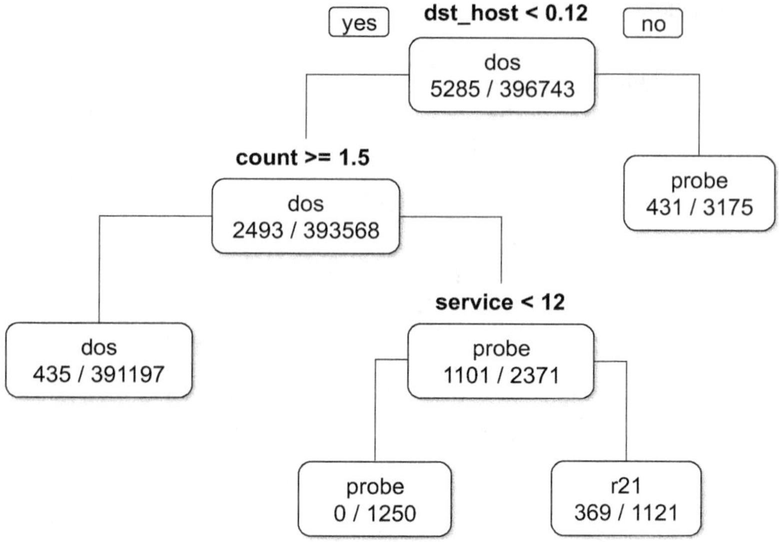

Fig. 3.26 Classify tree of abnormal traffic selection model

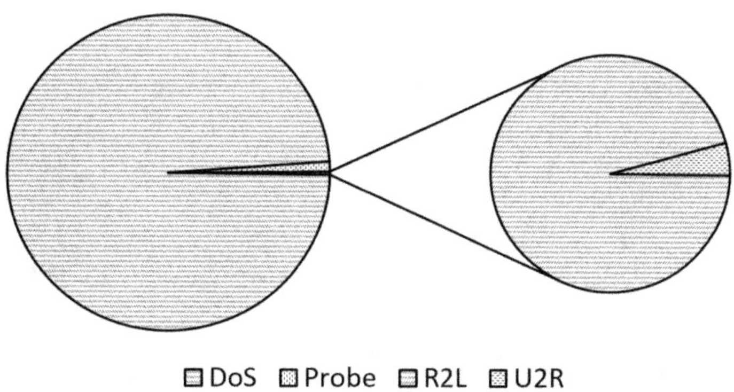

Fig. 3.27 Distribution of training data

Table 3.25 Accuracy of No. 3 and No. 4

No.	Model	Algorithm	Accuracy
No. 3	Normal traffic selection	k-means	0.926
	Abnormal traffic selection	Random forest	0.948
	Mixed compensation model		0.923
No. 4	Normal traffic selection	k-means	0.925
	Abnormal traffic selection	Decision tree	0.961
	Mixed compensation model		0.918

Table 3.26 Confusion matrix of normal traffic selection model of No. 3 and No. 4

No.	Prediction	Normal	Abnormal
No. 3	Normal	59,189	21,663
	Abnormal	1404	228,773
No. 4	Normal	59,428	22,221
	Abnormal	1165	228,215

Table 3.27 Results of experiments

No.	Experiment	DoS	Probe	U2R	R2L	Final	Rank
1	k-means1+Random forest	10	1	1	2	14	1
6	Decision tree+ Random forest	1	8	2	5	16	2
8	Random forest + Random forest	2	5	4	7	18	3
11	Random forest	2	4	3	9	18	3
3	k-means2+ Random forest	5	2	5	7	19	5
2	k-means1+ Decision tree	11	3	6	1	21	6
7	Random forest + Decision tree	7	6	6	4	23	7
5	Decision tree + Decision tree	4	9	6	6	25	8
9	k-means	9	11	6	3	29	9
4	k-means2+ Decision tree	8	7	6	10	31	10
10	Decision tree	5	10	6	11	32	11

Nowadays, many novel attacks are unknown to researchers, and many attacks will be disguised as normal. It's very dangerous to have high false negative rate, and it does not fit the proposed model.

Because the effect of k-means has great correlation with the number of centers chosen to cluster, and we can fine tune the strength of clustering, and lower the false negative rate to establish a strict normal selection model.

In No. 3 and No. 4, the number of centers for normal traffic and attacks is 100 and 300, respectively. Although it can achieve a good overall accuracy, its false negative rate is higher than other model. However, according to Table 3.27, by choosing 4 and 30 in No. 1 and No. 2, it has lower false negative rate, and only classify four kinds of attacks. Besides, a strict normal detection model is established.

By adjusting the parameters and reducing false negative rate in No. 1 and No. 2, we can find that the rank has increased rapidly compared with No. 3 and No. 4. Especially, when K-means combines with random forest, it has a very high accuracy on Probe, U2R and R2L attack. Therefore, we can draw the conclusion that by adjusting the parameters of K-means, the strength of abnormal traffic detection can be controlled by adjusting the strength of normal traffic identification.

Table 3.28 Summary of model

	Model 1	Model 2	Model 3
Normal traffic selection model	k-means1	Decision tree	Random forest
Abnormal traffic selection model	Random forest	Random forest	Random forest

Based on the results analyzed above, as shown in Table 3.28, the following conclusions can be drawn:

1. Random forest classification algorithm can adapt to the change of distribution of network data, and this algorithm by using the proposed model can reduce false negative rate.
2. If the number of training data in different group is largely different with each other, the classify model built by decision tree will prefer to attack types, which have more training data. So we should avoid using decision tree in abnormal traffic selection model. However, in the normal traffic selection model, the difference between different groups is comparatively small. In this situation, using decision tree can fast get classify model, and the results have higher accuracy.
3. There are more and more unknown abnormal events in the future. In order to avoid loss of false negative prediction, we can change the number of clustering in the normal traffic selection model with k-means algorithm to reduce false negative rate and increase the accuracy of detecting abnormal events.

3.4 Summary

In this chapter, we discuss the main challenge of network traffic intelligent awareness and introduced several machine learning based traffic awareness algorithms. we first present a multi-level intrusion detection model framework named MSML to address these issues. Then, we propose a novel IDS framework called HMLD to address these issues, which is an exquisite designed framework based on Hybrid Multi-Level Data Mining. In addition, we propose a new model based on big data analysis, which can avoid the influence brought by adjustment of network traffic distribution, increase detection accuracy and reduce the false negative rate. Finally, we proposes an end-to-end IoT traffic classification method relying on a deep learning aided capsule network for the sake of forming an efficient classification mechanism that integrates feature extraction, feature selection and classification model.

References

1. J. Du, C. Jiang, H. Zhang, X. Wang, and M. Debbah, "Secure satellite-terrestrial transmission over incumbent terrestrial networks via cooperative beamforming," *IEEE Journal on Selected Areas in Communications*, vol. PP, no. 99, pp. 1–1, 2018.
2. C.-H. Tsang, S. Kwong, and H. Wang, "Genetic-fuzzy rule mining approach and evaluation of feature selection techniques for anomaly intrusion detection," *Pattern Recognition*, vol. 40, no. 9, pp. 2373–2391, 2007. [Online]. Available: http://www.sciencedirect.com/science/article/pii/S0031320306005218
3. J. Wang, C. Jiang, H. Zhu, R. Yong, and L. Hanzo, "Internet of vehicles: Sensing aided transportation information collection and diffusion," *IEEE Transactions on Vehicular Technology*, vol. 67, no. 5, pp. 3813–3825, 2018.
4. J. Wang, C. Jiang, H. Zhang, Z. Xin, V. C. M. Leung, and L. Hanzo, "Learning-aided network association for hybrid indoor lifi-wifi systems," *IEEE Transactions on Vehicular Technology*, vol. PP, no. 99, pp. 1–1, 2017.
5. J. Du, C. Jiang, K. C. Chen, R. Yong, H. V. Poor, J. Du, C. Jiang, K. C. Chen, R. Yong, and H. V. Poor, "Community-structured evolutionary game for privacy protection in social networks," *IEEE Transactions on Information Forensics & Security*, vol. 13, no. 3, pp. 574–589, 2018.
6. X. Zhu, C. Jiang, L. Yin, L. Kuang, G. Ning, and J. Lu, "Cooperative multigroup multicast transmission in integrated terrestrial-satellite networks," *IEEE Journal on Selected Areas in Communications*, vol. PP, no. 99, pp. 1–1, 2018.
7. J. Erman, A. Mahanti, M. Arlitt, I. Cohen, and C. Williamson, Offline/realtime traffic classification using semi-supervised learning, *Performance Evaluation*, vol. 64, no. 9–12, pp. 1194–1213, 2007.
8. H. Yao, D. Fu, P. Zhang, M. Li, and Y. Liu, "Msml: A novel multi-level semi-supervised machine learning framework for intrusion detection system," *IEEE Internet of Things Journal*, 2018.
9. J. Zhang, X. Chen, Y. Xiang, W. Zhou, and J. Wu, "Robust network traffic classification," *IEEE/ACM Transactions on Networking*, vol. 23, no. 4, pp. 1257–1270, 2015.
10. J. Davis and M. Goadrich, "The relationship between precision-recall and roc curves," in *International Conference on Machine Learning*, 2006, pp. 233–240.
11. L. He, "An improved intrusion detection based on neural network and fuzzy algorithm," *Journal of Networks,9,5(2014-05-08)*, vol. 9, no. 5, 2014.
12. W. Xuren, H. Famei, and X. Rongsheng, "Modeling intrusion detection system by discovering association rule in rough set theory framework," in *International Conference on Computational Intelligence for Modelling, Control & Automation*, 2006, p. 24.
13. W. L. Al-Yaseen, Z. A. Othman, M. Z. A. Nazri, Multi-level hybrid support vector machine and extreme learning machine based on modified K-means for intrusion detection system, Expert Systems with Applications 67 (2017) 296–303.
14. KDD Cup 1999 Data, http://kdd.ics.uci.edu/databases/kddcup99/kddcup99.html.
15. Q. M. Alriyami, E. Asimakopoulou, N. Bessis, A Survey of Intrusion Detection Systems for Mobile Ad Hoc Networks, in: International Conference on Intelligent Networking and Collaborative Systems(INCoS), Salerno, Italy, 2014.
16. H. Yao, L. Chong, P. Zhang, and L. Wang, "A feature selection method based on synonym merging in text classification system," *Eurasip Journal on Wireless Communications & Networking*, vol. 2017, no. 1, p. 166, 2017.
17. D. E. Denning, An Intrusion-Detection Model, in: IEEE Symposium on Security and Privacy, Oakland, CA, USA, 1986.
18. Y. He, F. R. Yu, N. Zhao, H. Yin, H. Yao, R. C. Qiu, Big Data Analytics in Mobile Cellular Networks, IEEE Access 4 (2016) 1985–1996.
19. V. Vapnik, C. Cortes, Support vector networks, Machine Learning 20 (3) (1995) 273–297.

20. G. Zhang, B. E. Patuwo, M. Y. Hu, Forecasting with artificial neural networks:: The state of the art, International Journal of Forecasting 14 (1) (1998) 35–62.
21. J. R. Quinlan, Induction of Decision Trees, Machine Learning 1 (1) (1986) 81–106.
22. A. Cutler, D. R. Cutler, J. R. Stevens, Random Forests, Machine Learning 45 (1) (2004) 157–176.
23. A. K. Jain, M. N. Murty, P. J. Flynn, Data clustering: a review, ACM Computing Surveys(CSUR) 31 (3) (1999) 264–323.
24. J. A. Hartigan, M. A. Wong, A K-means Clustering Algorithm, Applied Statistics 28 (1) (1979) 100–108.
25. P. Zhang, S. Wu, M. Wang, H. Yao, and Y. Liu, "Topology based reliable virtual network embedding from a qoe perspective," *China Communications*, vol. 15, no. 10, pp. 38–50, 2018.
26. K. Khan, S. U. Rehman, K. Aziz, S. Fong, S. Sarasvady, DBSCAN: Past, present and future, in: Fifth International Conference on the Applications of Digital Information and Web Technologies (ICADIWT), Bangalore, India, 2014.
27. K. Wang, J. Zhang, D. Li, X. Zhang, T. Guo, Adaptive Affinity Propagation Clustering, Acta Automatica Sinica 33 (12) (2007) 1242–1246.
28. W. Wu, H. Yao, T. Huang, L. Wang, Y. Zhang, and Y. Liu, "Survey of development on future networks and industrial internet," *Journal of Beijing University of Technology*, vol. 43, no. 2, pp. 163–172, 2017.
29. G. Wang, J. Hao, J. Ma, L. Huang, A new approach to intrusion detection using Artificial Neural Networks and fuzzy clustering, Expert Systems with Applications 37 (9) (2010) 6225–6232.
30. P. Gogoi, D. K. Bhattacharyya, B. Borah, J. K. Kalita, MLH-IDS: A Multi-Level Hybrid Intrusion Detection Method, Computer Journal 57 (4) (2014) 602–623.
31. X. Zhu, C. Jiang, L. Kuang, G. Ning, and J. Lu, "Non-orthogonal multiple access based integrated terrestrial-satellite networks," *IEEE Journal on Selected Areas in Communications*, vol. PP, no. 99, pp. 1–1, 2017.
32. I. Guyon, A. Elisseeff, An Introduction to Variable and Feature Selection, Journal of Machine Learning Research 3 (6) (2003) 1157–1182.
33. P. Zhang, H. Yao, M. Li, and Y. Liu, "Virtual network embedding based on modified genetic algorithm," *Peer-to-Peer Networking and Applications*, no. 2, pp. 1–12, 2017.
34. M. Ambusaidi, X. He, P. Nanda, Z. Tan, Building an Intrusion Detection System Using a Filter-Based Feature Selection Algorithm, IEEE Transactions on Computers 65 (10) (2016) 2986–2998.
35. S. J. Horng, M. Y. Su, Y. H. Chen, T. W. Kao, R. J. Chen, J. L. Lai, C. D. Perkasa, A novel intrusion detection system based on hierarchical clustering and support vector machines, Expert Systems with Applications 38 (1) (2011) 306–313.
36. C. Jiang, C. Jiang, N. C. Beaulieu, L. Yong, Y. Zou, and R. Yong, "Dywamit: Asynchronous wideband dynamic spectrum sensing and access system," *IEEE Systems Journal*, vol. 11, no. 3, pp. 1777–17882, 2017.
37. C. Elkan, Results of the KDD'99 classifier learning, ACM SIGKDD Explorations Newsletter 1 (2) (2000) 63–64.
38. Patcha, A.; Park, J.M. (2007); An overview of anomaly detection techniques: Existing solutions and latest technological trends, *Computer Networks*, ISSN 1389-1286, 51(12): 3448–3470.
39. L. Kuang, C. Xi, C. Jiang, H. Zhang, and W. Sheng, "Radio resource management in future terrestrial-satellite communication networks," *IEEE Wireless Communications*, vol. 24, no. 5, pp. 81–87, 2017.
40. J. Wang, C. Jiang, H. Zhu, R. Yong, and L. Hanzo, "Taking drones to the next level: Cooperative distributed unmanned-aerial-vehicular networks for small and mini drones," *IEEE Vehicular Technology Magazine*, vol. 12, no. 3, pp. 73–82, 2017.
41. Lazarevic, A.; Kumar, V.; Srivastava, J. (2005); Intrusion detection: A survey, *Managing Cyber Threats*, ISSN 0924-6703, 5: 19–78.

42. Axelsson, S. (1998); Research in intrusion-detection systems: a survey, *Department of Computer Engineering, Chalmers University of Technology, Goteborg. Sweden*, Technical Report 98-17.
43. H. Yao, T. Mai, X. Xu, P. Zhang, M. Li, and Y. Liu, "Networkai: An intelligent network architecture for self-learning control strategies in software defined networks," *IEEE Internet of Things Journal*, vol. 5, no. 6, pp. 4319–4327.
44. Om, H.; Kundu, A. (2012); A hybrid system for reducing the false alarm rate of anomaly intrusion detection system, *IEEE 1st International Conference on Recent Advances in Information Technology (RAIT)*, ISBN 978-1-4577-0694-3, 131–136.
45. Kaisler, S. et al (2013); Big data: Issues and challenges moving forward, *IEEE 46th Hawaii International Conference on System Sciences (HICSS)*, ISSN 1530-1605, 995–1004.
46. Michael, K.; Miller, K.W. (2013); Big Data: New Opportunities and New Challenges, *Computer*, ISSN 0018-9162, 46(6):22–24.
47. P. Zhang, H. Yao, and Y. Liu, "Virtual network embedding based on the degree and clustering coefficient information," *IEEE Access*, vol. 4, pp. 8572–8580
48. Russom, P. et al (2011); Big Data Analytics, *TDWI Best Practices Report*, Fourth Quarter.
49. J. Du, E. Gelenbe, C. Jiang, H. Zhang, Y. Ren, J. Du, E. Gelenbe, C. Jiang, H. Zhang, and Y. Ren, "Contract design for traffic offloading and resource allocation in heterogeneous ultra-dense networks," *IEEE Journal on Selected Areas in Communications*, vol. 35, no. 11, pp. 2457–2467, 2017.
50. Fan, W.; Bifet, A. (2013); Mining big data: current status, and forecast to the future, *ACM SIGKDD Explorations Newsletter*, ISSN 1931-0145, 14(2): 1–5.
51. James, G. et al (2013); An introduction to statistical learning, *Springer*, ISSN 1431-875X.
52. H. Yao, L. Wang, X. Wang, L. Zhou, and Y. Liu, "The space-terrestrial integrated network (stin): An overview," *IEEE Communications Magazine*, vol. 56, no. 9, pp. 178–185, 2018.
53. KDD Cup 1999, http://kdd.ics.uci.edu/databases/kddcup99/kddcup99.html. May 2015
54. Lippmann, R.P. et al (2000); Evaluating intrusion detection systems: The 1998 DARPA off-line intrusion detection evaluation, *IEEE Proceedings of DARPA Information Survivability Conference and Exposition (DISCEX)*, ISBN 0-7695-0490-6, 2:12–26.
55. B. Deng, C. Jiang, L. Kuang, G. Song, J. Lu, and S. Zhao, "Two-phase task scheduling in data relay satellite systems," *IEEE Transactions on Vehicular Technology*, vol. 67, no. 2, pp. 1782–1793, 2017.
56. H. Yao, C. Xu, M. Li, P. Zhang, and L. Wang, "A novel virtual network embedding algorithm based on policy network and reinforcement learning," *Neurocomputing*, vol. 284, pp. 1–9, 2018.
57. Tavallaee, M. et al (2009); A detailed analysis of the KDD CUP 99 data set, *Proceedings of the Second IEEE Symposium on Computational Intelligence for Security and Defence Applications (CISDA)*, ISBN 978-1-4244-3763-4, 1–6.
58. P. Zhang, H. Yao, C. Qiu, and Y. Liu, "Virtual network embedding using node multiple metrics based on simplified electre method," *IEEE Access*, vol. 6, pp. 37 314–37 327, 2018.
59. Pfahringer, B. (2000); Winning the KDD99 classification cup: bagged boosting, *ACM SIGKDD Explorations Newsletter*, ISSN 1931-0145, 1(2): 65–66.
60. Yu, G. D. et al (2014); Multi-objective rescheduling model for product collaborative design considering disturbance, *International journal of simulation modelling*, ISSN 1726-4529, 13(4): 472–484.
61. Gusel, L. R. et al (2015); Genetic based approach to predicting the elongation of drawn alloy, *International journal of simulation modelling*, ISSN 1726-4529, 14(1): 39–47.
62. J. Wang, C. Jiang, Z. Kai, T. Q. S. Quek, R. Yong, and L. Hanzo, "Vehicular sensing networks in a smart city: Principles, technologies and applications," *IEEE Wireless Communications*, vol. 25, no. 1, pp. 122–132, 2018.
63. Prasad, K. et al (2016); A knowledge-based system for end mill selection, *Advances in Production Engineering & Management*, ISSN 1856-6250, 11(1): 15–28.

Chapter 4
Intelligent Network Control

Finding the near-optimal control strategy is the most critical and ubiquitous problem in a network. Examples include routing decision, load balancing, QoS-enable load scheduling, and so on. However, the majority solutions of these problems are largely relying on a manual process. To address this issue, in this chapter, we apply several artificial intelligence approaches for self-learning control strategies in networks. In this chapter, we first present an energy-aware multi-controller placement scheme as well as a latency-aware resource management model for the SDWN. Moreover, the particle swarm optimization (PSO) is invoked for solving the multi-controller placement problem, and a deep reinforcement learning (DRL) algorithm aided resource allocation strategy is conceived. Then, we present a novel controller mind (CM) framework to implement automatic management among multiple controllers and propose a novel Quality of Service (QoS) enabled load scheduling algorithm based on reinforcement learning to solve the problem of complexity and pre-strategy in the networks. In addition, we present a Wireless Local Area Networks (WLAN) interference self-optimization method based on a Self-Organizing Feature Map (SOM) neural network model to suppress the interference in local area networks. Finally, we propose a BC-based consensus protocol in distributed SDIIoT, where BC works as a trusted third party to collect and synchronize network-wide views between different SDN controllers. In addition, we use a novel dueling deep Q-learning approach to solve this joint problem.

4.1 Multi-Controller Optimization in SDN

Given the proliferation of mobile data traffic in a variety of wireless networks [1], traditional network resource management techniques may not satisfy the users' quality of experience (QoE). The software defined wireless networks (SDWN) [2] was proposed with the spirit of decoupling the control layer and the infrastructure

H. Yao et al., *Developing Networks using Artificial Intelligence*, Wireless Networks, https://doi.org/10.1007/978-3-030-15028-0_4

layer, which can both efficiently support big data driven resource management and beneficially provide a global optimal system performance.

In retrospect, how to appropriately deploy controllers in different positions has become a critical problem for wired software defined networks (SDNs), also termed as the controller placement problem (CPP) [5], which attracts numerous heuristic algorithms to solve it. By contrast, few works focused their attention on the CPP in the context of SDWN architecture. Recently, Abdel et al.[6] investigated the CPP between the controller and the elements under its control considering average response time and maximum response time. Resource management is another key problem, since traditional resource management methods cannot support online and adaptive decision making corresponding to the real-world dynamic experimental states. Hence, Grandl et al.[8] proposed a Tetris assisted resource scheduling framework and Jalaparti et al.[10] designed a Corral aided framework, where they defined the resource management as a combinatorial optimization problem, and both frameworks yielded an improved performance. Moreover, given the successful applications of deep learning algorithms in image processing and robotics, Mao et al.[12] utilized deep reinforcement learning (DRL) algorithms for adaptively solving resource allocation problems [25].

However, most of them only focused on the latency for multi-controller placement without taking into account the impact of energy consumption imposed on the CPP, where controllers may be energy constrained. Furthermore, resource management of each controller in SDWNs substantially influence the QoE of users such as the waiting time. Inspired by aforementioned issues, in this part, we propose an energy-aware multi-controller placement scheme as well as a DRL aided resource management model for the SDWN. The main contributions of our paper can be summarized as follows.

- A particle swarm optimization (PSO) aided multi-controller placement model is conceived for minimizing the system's energy consumption with latency constraints. Relying on PSO, our proposed energy-aware multi-controller placement scheme can be efficiently solved.
- Relying on the powerful expression capability of deep reinforcement learning algorithms, we propose a DRL based resource management scheme for the SDWN, where the self-aware controller system is capable of reducing the waiting time of each task.

4.1.1 *System Model*

4.1.1.1 Network Model

In our model, the SDWN contains n controllers and m elements. Let $\mathbf{C} = \{c_1, c_2, \cdots, c_n\}$ denote the set of the controllers and $\mathbf{S} = \{s_1, s_2, \cdots, s_m\}$ be the set of the elements. Elements and controllers are capable of communicating with each other via wireless links. And the maximum number of the elements that a single controller can support is represented by L. Let m_i be the number of elements served by controller c_i, and we have $m_i \in (0, n-i+1]$ and $\sum_{i=1}^{n} m_i = m$. Moreover, the distance between c_i and one of its clients, i.e. s_{i,m_i}, is denoted as $d(c_i, s_{i,m_i})$. Task j is represented by $\Omega_j = (w_j, \eta_j)$, where w_j denotes the amount of computation of Ω_j, that is the required CPU cycles for completing the task, while η_j represents the amount of communication traffic in Ω_j, i.e. the amount of the data transmitted toward the element.

4.1.1.2 Communication Model

In our communication model, $h(c_i, s_{i,m_i})$ represents the channel gain between controller c_i, $(i = 1, 2, \cdots, n,\ c_i \in \mathbf{C})$ and the element s_{i,m_i} served by c_i. Let $p(c_i)$ represent the transmission power of controller c_i. Hence, the uplink data rate $r(c_i, s_{i,m_i})$ of this link can be calculated as:

$$r_i(c_i, s_{i,m_i}) = B \log_2 \left(1 + \frac{p(c_i)h(c_i, s_{i,m_i})}{\sigma^2 + I(c_i, s_{i,m_i})}\right), \tag{4.1}$$

where σ^2 is the variance of the white Gaussian noise and B denotes the channel bandwidth, while $I(c_i, s_{i,m_i})$ represents the inter-cell interference between controller c_i and element s_{i,m_i}. The *transmission delay* from the controller c_i to the element s_{i,m_i} for task Ω_j can be formulated as:

$$t_i^T(c_i, s_{i,m_i}) = \frac{\eta_j}{r(c_i, s_{i,m_i})}. \tag{4.2}$$

As for the *transmission energy consumption* of controller c_i, we have:

$$\varepsilon^T(c_i, s_{i,m_i}) = p(c_i)t^T(c_i, s_{i,m_i}) = \frac{p(c_i)\eta_j}{r(c_i, s_{i,m_i})}. \tag{4.3}$$

4.1.1.3 Computation Model

Let us define f_i as the computational capacity of the controller c_i. Hence, the controller's *execution time* of task $\Omega_j = (w_j, \eta_j)$ can be expressed as:

$$t_i^C(c_i, s_{i,m_i}) = \frac{w_j}{f_i}. \tag{4.4}$$

Furthermore, we can obtain the *execution energy consumption* of controller c_i for processing task Ω_j as:

$$\varepsilon_i^C(c_i, s_{i,m_i}) = \rho_i w_j, \tag{4.5}$$

where ρ_i is the unit execution energy consumption of one CPU cycle in the controller.

4.1.1.4 Problem Formulation

Considering both the limited computational capacity and the energy constraint of controllers, we propose an energy-aware multi-controller placement scheme as well as a latency-aware resource management model for the SDWN. Our objective is to minimize the average energy consumption of all controllers for both communication and computation in large-scale SDWNs. This placement problem can be formulated as follows:

$$\begin{aligned}
&\min \frac{1}{n}\sum_{i=1}^{n}\left[\varepsilon_i^T(c_i, s_{i,m_i}) + \varepsilon_i^C(c_i, s_{i,m_i})\right] \\
&\text{s.t.} \quad (4.6\text{-a}): t_i^T(c_i, s_{i,m_i}) + t_i^C(c_i, s_{i,m_i}) \le t_{\max}, c_i \in \mathbf{C}, s_{i,m_i} \in \mathbf{S}, \\
&\qquad (4.6\text{-b}): d(c_i, s_{i,m_i}) \le d_{\max}, c_i \in \mathbf{C}, s_{i,m_i} \in \mathbf{S}, \\
&\qquad (4.6\text{-c}): m_i \le L(c_i), c_i \in \mathbf{C} \\
&\qquad (4.6\text{-d}): \sum_{i=1}^{n} m_i = m,
\end{aligned} \tag{4.6}$$

where $t_{\max}$ in (4.6-a) represents the maximum time requirement of the task, while $d_{\max}$ in (4.6-b) is the maximum deployment distance between c_i and s_{i,m_i}. $L(c_i)$ in (4.6-c) represents the maximum number of the elements that the controller c_i can support. In Sect. 4.1.2, PSO algorithm will be invoked for solving our proposed multi-controller placement problem.

After obtaining a delicate multi-controller placement strategy from (4.6), we try to further determine computational resource allocation scheme of the controller for each task. Hereinafter, DRL algorithm is employed to solve this problem. For the task Ω_j, its *ideal completion time* is defined by $T_{\text{ideal},\Omega_j}$, while the *actual execution time* is defined by $T_{\text{actual},\Omega_j}$. Hence, we aims for minimizing the its *waiting time* Γ_{Ω_j}, i.e.

$$\min \quad \Gamma_{\Omega_j} = T_{\text{actual},\Omega_j} - T_{\text{ideal},\Omega_j}. \tag{4.7}$$

4.1.2 *Methodology*

4.1.2.1 PSO Aided Near-Optimal Multi-Controller Placement

The PSO is a family member of stochastic optimization algorithms, which is inspired by the social behavior of birds [45]. In PSO, for the i-th particle in the d-dimensional solution space, its position can be represented as $X_i = (x_{i1}, x_{i2}, \cdots, x_{id})$ and its velocity can be represented as $V_i = (v_{i1}, v_{i2}, \cdots, v_{id})$. Relying on the fitness function, the individual best position of particle i is expressed as $P_i = (p_{i1}, p_{i2}, \cdots, p_{id})$, while its global best position is $P_g = (p_{g1}, p_{g2}, \cdots, p_{gd})$. Hence, the velocity in the iteration $(n + 1)$ can be updated by:

$$V_i(n + 1) = \omega V_i(n) + c_1\xi_1(P_i - X_i) + c_2\xi_2(P_g - X_i), \tag{4.8}$$

where c_1 and c_2 are two positive constants generated in $[0, 2]$, while ξ_1 and ξ_2 are a pair of independent random numbers valued in $(0, 2)$. ω is a weight coefficient for the sake of ensuring the convergence of PSO algorithm. Based on the updated velocity, particle's position can be given by:

$$X_i(n + 1) = X_i(n) + V_i(n + 1). \tag{4.9}$$

In this part, we apply PSO algorithm to solve the multi-controller placement problem, where particles are represented as the possible position of the controllers. After iterations, a near-optimal solution of controller placement can be obtained, as shown in Algorithm 1.

4.1.2.2 Resource Management Relying on Deep Q-Learning

In the following, we study the computational resource allocation scheme of the controller for each task relying on DRL algorithm [4, 11]. In the process of interacting with the environment, the model-free Q-learning algorithm is capable of finding the near-optimal behavior by continuous trial and error. Assume the state in the step n is presented as S_n, π means the policy space, and the accumulated reward $V^\pi(S_n)$ of state S_n is calculated as:

$$V^\pi(S_n) = R_n + \gamma R_{n+1} + \gamma^2 R_{n+2} + \cdots, \tag{4.10}$$

where R_n is the instant reward in step n, while $\gamma (0 < \gamma < 1)$ represents the discount factor which reflects the influence of the future reward. More explicitly, a large γ means that the training agent focuses more on to the future experience, otherwise the training agent only concerns about the immediate reward [55]. In Q-learning algorithm, the Q-value is the function of state S and action A, which is formulated as:

$$Q(S_n, A_n) = R_n + \gamma V^\pi(S_{n+1}). \tag{4.11}$$

Algorithm 1 PSO aided multi-controller placement

Input: network topology, objective function, the number of controllers m, the number of particles s;
Output: best controller placement x^*;
Initialization:
Randomly generate particle's position X_i and velocity V_i;
Calculate the fitness value of each particle according to objective function;
Obtain the individual best position P_i and the global best position P_g.

while $i < n$ **do**
  for each particle X_i
Update particle's velocity V_i according to Eq. (4.8);
Update particle's position X_i according to Eq. (4.9);
Calculate the fitness value of particle $Fitness(X_i)$;

  if $Fitness(X_i) < Fitness(P_i)$ **then**
    Update the individual best position P_i;

  end if
  if $Fitness(X_i) < Fitness(P_g)$ **then**
    Update global best position P_g;

  end if
end while

Thus, $Q_{n+1}(S_n, A_n)$ can be updated by:

$$\begin{aligned} Q_{n+1}(S_n, A_n) &= (1-\eta)Q_n(S_n, A_n) \\ &\quad + \eta[R_n + \gamma \max Q_n(S_{n+1}, A_{n+1})], \end{aligned} \tag{4.12}$$

where the learning rate $0 < \eta < 1$ controls the learning convergence speed.

Deep Q-network utilizes a feedforward artificial neural network for the sake of approximating the Q-value function, i.e $Q(S_n, A_n; \theta)$. The deep Q-network is trained in the form of minimizing a loss function by updating θ in terms of small steps. The loss function can be given by:

$$L(\theta) = E[(y(S_n, A_n, S_{n+1}; \hat{\theta}) - Q(S_n, A_n; \theta))^2], \tag{4.13}$$

where the target function $y(S_n, A_n, S_{n+1}; \hat{\theta}) = R + \gamma \max Q(S_{n+1}, A_{n+1}; \hat{\theta})$.

In this part, we apply the deep Q-network algorithm to solving the resource management problem, as shown in Algorithm 2. For each SDWN controller, we consider there are total Ω tasks waiting for service. Three key factors of the deep Q-network are given as follows.

- *State Space*: The available resources of controllers includes the current allocated computational resources and the required computational resources of the tasks which may be scheduled [62]. We define the demand of computational resources of the task j with type z by $u_{j,z}$, and the current computational resource level of the controller i as $u_{i,z}$. Hence, considering a total of Z kinds of tasks, the system's state with total m controllers can be given by $\Lambda = [S_1, S_2, \cdots, S_m]$, where $S_m = [u_{j,k}, u_{m,k}, z = 1, 2, \cdots, Z]$.

- *Action Space*: The serving queue can hold maximum K tasks and each controller is capable of dealing with only one task in each time slot. For the controller m, let the action $A_m = k$ represent the k-th task is scheduled at the current time slot, while $A_m = \phi$ means that no task is scheduled in the current time slot. Then the action space of controller m can be expressed as $\{\phi, 1, 2, \cdots, K\}$.
- *Reward*: The reward function is designed for directing the agent to minimize the average waiting time in the controllers. Specifically, we set the reward at each time slot to $\sum 1/T_{\text{ideal},\Omega_j}$, $\Omega_j \in \Omega$, where Ω is the set of current tasks supported by the controller.

Algorithm 2 Deep Q-network based resource management strategy

Input: state S, action A and the maximum number of iterations N
Output: $Q(S, A)$;
Initialization:
Initialize the weight of deep Q-network θ;

for $n < N$ **do**
  if random probability $p < \delta$ **then**
    select an action A_n;
otherwise
$A_n = \arg\max Q(S_n, A_n; \theta)$
  end if
Execute action A_n, then obtain the reward R_n and arrive at the next state S_{n+1};
Calculate the target Q-value
$y(S_n, A_n, S_{n+1}; \hat{\theta}) = R + \gamma \max Q(S_{n+1}, A_{n+1}; \hat{\theta})$
Update the deep Q-network by minimizing the loss $L(\theta)$;
$L(\theta) = E[(y(S_n, A_n, S_{n+1}; \hat{\theta}) - Q(S_n, A_n; \theta))^2]$
end for

4.1.3 Simulation Results

In this subsection, we evaluate proposed PSO algorithm relying on six different network topologies from the Internet zoo [7, 14]. The channel bandwidth is $B = 5\,\text{MHz}$ and the transmission power is $p = 50\,\text{mW}$. The variance of white Gaussian noise is $\sigma^2 = -100\,\text{dBm}$. The channel gain is set as $h = d^{-\lambda}$, where $\lambda = 4$ represents the path loss factor. The amount of one task is 20 Mb. $\rho = 4\,\text{J/GHz}$ denotes the unit energy consumption per CPU circle. The computational capacity of the controller is $f = 2\,\text{GHz}$. In order to verify the performance of our proposed algorithm, we use the genetic algorithm (GA) as well as the greedy algorithm (GL) for comparison. The number of particles is 20. Moreover, some parameters of proposed two algorithms are listed in Table 4.1.

Table 4.1 Parameters of two algorithms

Algorithm	Parameters
PSO	$c_1 = 2.0, c_2 = 2.0, w = [0.4, 0.9]$
GA	$P_c = 0.8, P_m = 0.1$

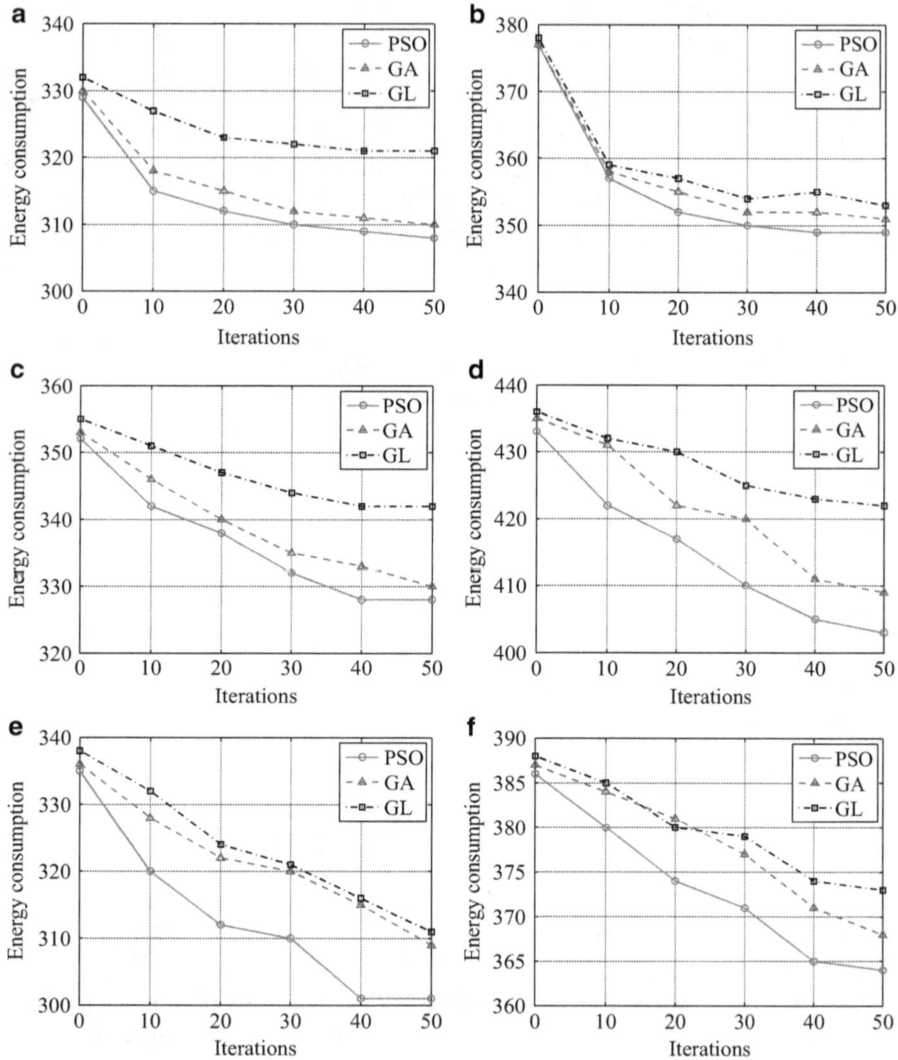

Fig. 4.1 Energy consumption versus iterations in six different network topologies (**a**) Aarnet. (**b**) Arn. (**c**) Bandcon. (**d**) Bellcanada. (**e**) Esnet. (**f**) ChinaNet

Figure 4.1 shows the performance of energy consumption versus iterations in six different topologies which assumed the link between nodes is wireless, i.e. Aarnet, Arn, Bandcon, Bellcanada, Esnet and ChinaNet. As shown in Fig. 4.1, with the increase of the iteration, PSO has the minimum energy consumption in comparison to the other two algorithms in the context of $m = 4$ controllers. We can conclude that the proposed PSO algorithm can provide a beneficial position placement scheme having the minimum energy consumption considering the latency constraints. By contrast, GL has the worst performance because it is easy to be trapped into local optimum. Hence, simulation results show that the PSO algorithm can obtain the near-optimal solution of controller placement for SDWNs.

As for the resource allocation, we use a neural network having a fully connected hidden layer with 30 neurons to construct our deep Q-learning network. The network's parameters are updated with learning rate 0.002. There are total three kinds of tasks, and the serving queue can hold at most 20 tasks. Moreover, the *ideal completion time* of each task is randomly generated from [1,10]. The simulation results are shown in Fig. 4.2. In Fig. 4.2a, we show the average waiting time versus the controller's load compare with both the short job first (SJF) algorithm and random serving scheme. We can conclude that with the increase of controller's load, our proposed DRL algorithm outperforms the others, which can improve the efficiency of the controllers. Moreover, as shown in Fig. 4.2b, a large

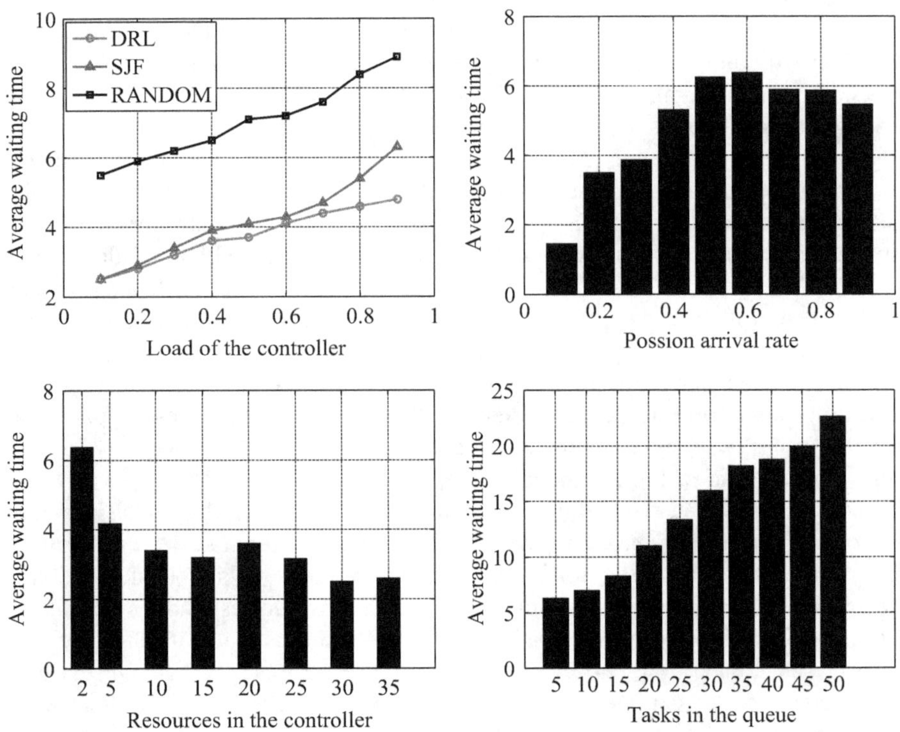

Fig. 4.2 Average waiting time versus different parameters. (**a**) Load in the controller. (**b**) Task arrival rate. (**c**) Resources in the controller. (**d**) Queue length

Poisson arrival rate of tasks process increases the average waiting time of the system. In Fig. 4.2c, it can be seen that the average waiting time of the system is reduced with the increase of computational resources of the controllers. Finally, Fig. 4.2d shows that the number of tasks in the queue increases the average waiting time of the system.

4.2 QoS-Enabled Load Scheduling Based on Reinforcement Learning

Energe resources crisis and global warming have become two global concerns [92]. As reasonable solutions, smart grid [16] and Energy Internet (EI) are seen as a new generation of energy provision paradigm, where improved communication mechanism is important to enable end-to-end communication. Software-defined networking (SDN) [18] is seen as a promising paradigm shift to reshape future network architecture, as well as smart grid and EI, called software-defined EI (SDEI). Using SDN enables to improve smart grid and EI by providing an abstraction of underlying network resources, forming global view for applications from upper layers, and decoupling infrastructures and control plane to enhance the flexibility and reliability of the system [21]. Noteworthy, the control plane is considered as the brain of SDN [22]. With the explosion of network scales and network traffic, overload in a single controller is one of the most intractable issues [118]. There is a growing consensus that the control plane should be designed as a multiple controllers plane to constitute a logically centralized but physically distributed model [27, 28, 32]. So far, the issues of multiple controllers have been studied in literature. Except for addressing the consistency problem of global view among distributed control plane, another key issue is how to schedule loads among multiple controllers so as to mitigate the risk of overloads and failures in one single controller.

On the other hand, the most important application of SDN in smart grid is real-time monitoring and communicating [96]. It follows that these applications require steady web-environment with no packet loss and less time delay to keep high accuracy and real time capability [33].

Traditionally, load scheduling algorithms make load scheduling decisions after the overload problems have happened [102]. In general, the traditional algorithms have three steps, including collecting load information, making load scheduling decisions, and sending load scheduling commands to the corresponding controllers. For example, the work in [34], load scheduling decision is made after the problem of overload. In addition, current CPU usage, current memory usage, current hard disk usage, and weight coefficient need to be exchanged among controllers when new load scheduling decision is made, which occupies lots of extra time so as to decrease time efficiency.

Recently, Machine learning (ML) has emerged as a novel technique, which can be used to tackle the above challenges [106]. Reinforcement Learning (RL) is regarded as an important branch of ML to solve complex control problems. RL is quite different from the traditional management and operation methods. It develops computer models to train datasets, which resolves the problems without being explicitly programmed to, but learning from environments.

4.2.1 System Description

In this subsection, we first give brief overviews of energy Internet and software-defined energy Internet. Then the CM framework in SDEI is presented.

4.2.1.1 Energy Internet

With energy crisis and limitation around world, how to use renewable energy has attracted lots of attentions, ranging from government and industry to academia [15]. Here, the development of renewable energy, as well as information and communication technologies (ICTs) are two key enablers of energy Internet [13, 19, 50]. Thus, energy Internet can be seen as an energy-utilizing system, combining distributed renewable energy with the advanced ICTs, which is known as the version of smart grids 2.0 [52].

Specially, ICTs provide a viable way to use the control capability of smart grid and allow distributed energy to access to the backbone grid in EI [9, 53]. Here, smart grid is used to collect and operate the information about the behaviors of users and suppliers to improve the sustainability and reliability of energy.

4.2.1.2 Software-Defined Energy Internet

With traditional TCP/IP protocol, energy Internet has achieved great success [111]. However, many challenges have emerged with the increasing number of smart connected devices in smart grid. It is hard for such rigid and static Internet to meet the demands of flexibility, agility, and ubiquitous accessibility.

In order to solve this embarrassment, there is a consensus to establish the future energy Internet architecture. SDN is seen as one of the most promising paradigms [54]. It is an approach to implement the network that separates the control plane and the data plane, abstracts the underlying infrastructures, and simplifies the network management by introducing the ability of programming [116].

Some works have employed SDN in energy Internet and smart grid [23]. For example, in order to support secure communications, the authors in [56] learned a SDN-enabled multi-attribute secure architecture for smart grid in IIoT environment. Moreover, the authors in [57] proposed a software defined advanced metering infrastructure (AMI) communication architecture. By this architecture, the problem of global load-balanced routing was solved.

Based on these works, we consider a software-defined energy Internet. However, before the wide adoption of SDEI, there are some problems remaining to be solved. The most intractable one is the scalability and reliability of control plane in SDEI. It can be anticipated that a logically centralized, but physically distributed control plane is necessary. Thus, we propose a controller mind framework in distributed SDEI to implement automatic management among multiple controllers.

4.2.1.3 Controller Mind framework

Based on traditional SDN architecture, we propose CM framework bridging the control plane and the data plane transparently, as shown in Fig. 4.3.

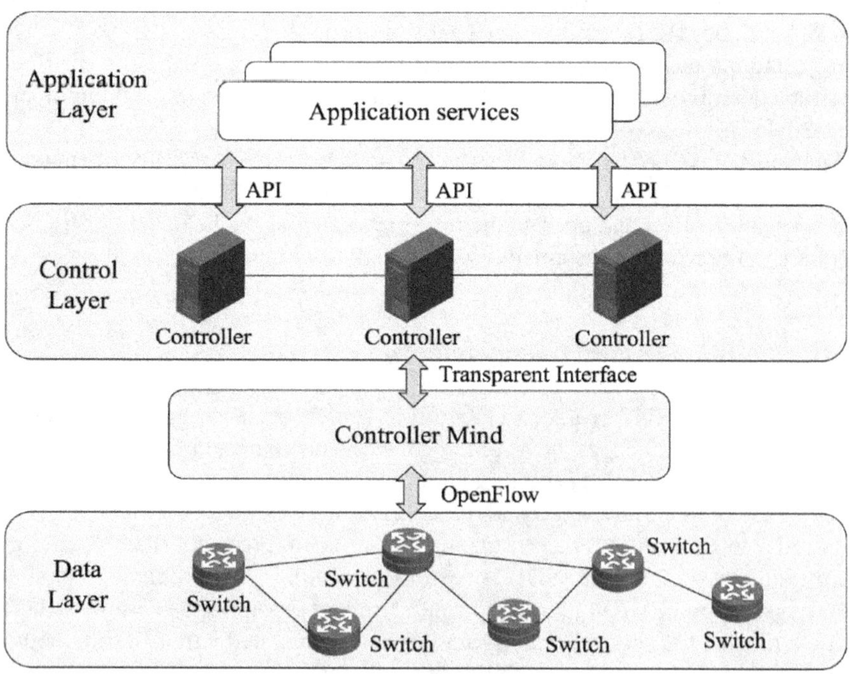

Fig. 4.3 A framework of controller mind in SDEI

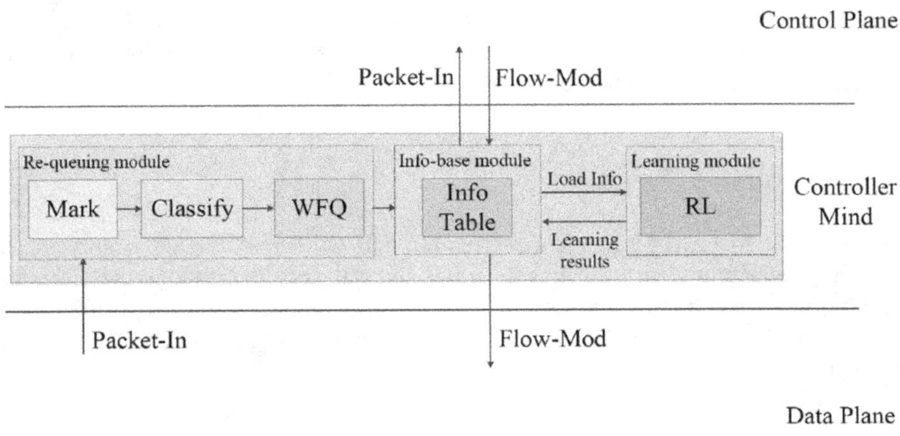

Fig. 4.4 The detailed CM framework of QoS-enabled load scheduling

The detailed structure of CM framework is given in Fig. 4.4. CM takes the responsibility of re-queuing incoming *Packet-in* flows to guarantee QoS and forwarding them to the appropriate controllers based on the results of reinforcement learning. Therefore, the CM framework consists of three modules, including *re-queuing module*, *info-base module* and *learning module*.

4.2.1.4 Re-Queuing Module

We assume that there are two types of traffic flows, namely QoS flows with high-priority and best-effort flows with low-priority. QoS flows include some traffic about the application of real-time monitoring in SDEI, and best-effort flows include some traffic from other applications with the low-level requirement of real-time capability [24]. When these traffic flows are encapsulated into *Packet-in* messages and sent to CM, based on source/destination MAC address, IP address, and TCP/UDP port in the packet headers, the re-queuing module marks and classifies the incoming *Packet-in* messages as QoS flows and best-effort flows [29, 35, 59], then re-queues them by the method shown in Sect. 4.2.2.1.

4.2.1.5 Info-Table Module

It receives re-queuing *Packet-in* messages from *re-queuing module*, sends them to controllers based on the learning results from the *learning module*, receives *Flow-mod* messages from the control plane, and sends *Flow-mod* messages to the data plane. The corresponding changes of the loads in each controller are aware by sending *Packet-in* messages, and receiving *Flow-mod* messages, which are all recorded by this module. On the one hand, these load records are the training datasets of *learning module*, on the other hand, by this mechanism, the frequent signaling interactions that are used to obtain the current load information are avoided, compared with the traditional schemes.

4.2.1.6 Learning Module

Based on the historical load records from the *info-table module* and reinforcement learning algorithm, *learning module* trains the data offline, obtains the learning results, and sends to *info-table module*. The reinforcement learning algorithm, i.e., Q-learning, is executed in this module, and the detail of the algorithm will be shown in Sect. 4.2.3.

4.2.2 System Model

In this subsection, we present re-queuing model, followed by workload model.

4.2.2.1 Re-Queuing Model

If we use *FIFO* (First In First Out) model, some delay-sensitive flows can't be treated fairly. Therefore, when arriving at CM, *Packet-in* messages would be classified into QoS messages or best-effort messages by extracting their headers, such as source/destination MAC address, IP address and TCP/UDP port. The architecture of classification and re-queuing at CM is shown in Fig. 4.5.

Here, we use a weight fair queuing(WFQ)[20, 60] algorithm in re-queuing model. In WFQ, we give weight coefficient w_i to classify queue i. Each classified queue sends the messages based on weight coefficient w_i. When the queue is empty, skip and access the next queue. Hence, the average service time of queue i is $\frac{w_i}{\sum w_j}$, where $\sum w_j$ is the sum of weight coefficients of all non-empty queues.

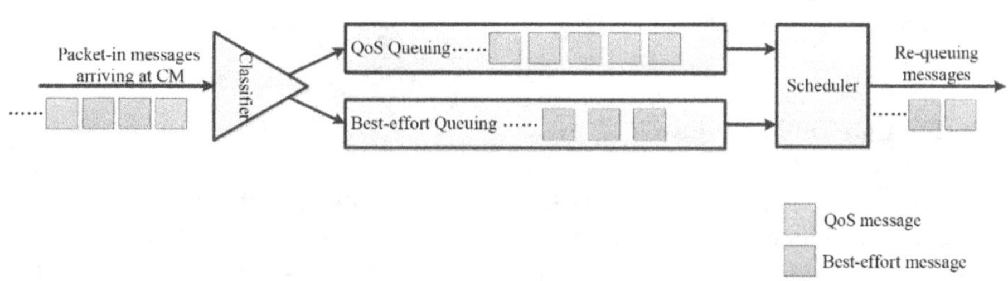

Fig. 4.5 Classification and re-queuing at CM

4.2.2.2 Workload Model

With the rapid development of commercial deployment, the performance of SDN controller is more and more important [36]. Global SDN certified testing center (SDNCTC) [61] develops a SDN controller performance test tool, called *OFsuite_Performance*, and releases a RYU controller performance test report [17, 63]. In this report, when the arriving rate of *Packet-in* messages is lower, the flow response time is linear with the *Packet-in* messages arriving rate, and the acceptable maximum arriving rate of *Packet-in* messages is related to the controller's own performance parameters. Hence, from the test result in the report, we obtain the relationship between the response time and the number of *Packet-in* messages in (4.14),

$$\tau = \rho N_p + \beta, N_p \leq max, \tag{4.14}$$

where τ represents the response time, and N_p is the number of *Packet-in* messages, ρ and β are the parameters related to the performance of each controller. max is the maximum number of *Packet-in* messages in the controller. This equation is an empirical expression that is shown in RYU controller performance test report. Although it has not ever been used in previously published work, we can have this relationship according to the real test works.

Meanwhile, Zhang et al. in [65] provided the relationship between the response time and the servers' load status as shown in (4.15),

$$\tau = \theta^{l_s}, \tag{4.15}$$

where l_s is the load status of servers. θ is a parameter related to server's performance. Controllers are always deployed in servers, so in this part we use the same load status model in controllers as servers.

From (4.14) and (4.15), we can deduce (4.16), which explains the relationship between the load status and the number of *Packet-in* messages in the controller.

$$l_s = log_\theta(\rho N_p + \beta), N_p \leq max. \tag{4.16}$$

When $N_p = max$, the load status is 100%. Thus using (4.16), we have $log_\theta(\rho max + \beta) = 1$, and there is a necessary relationship between the parameters, which is $\rho max + \beta = \theta$. Obviously, when $N_p > max$, the load status is also 100%. Thus we have (4.17),

$$l_s = \begin{cases} log_\theta(\rho N_p + \beta) & N_p \leq max \\ 100\% & N_p > max \end{cases}, \tag{4.17}$$

where $\rho max + \beta = \theta$

In the following problem formulation and simulation, we use (4.17) as the workload model. Since the number of *Packet-in* messages N_p is recorded by the info-table module, we use (4.17) to transfer the number of *Packet-in* messages to the load status of controllers to formulate and simulate the problem in the following sections.

4.2.3 *Problem Formulation*

In this subsection, we first describe a brief review of reinforcement learning. Then we formulate the load scheduling problem as a Q-learning process, which starts with the optimization problem formulation.

4.2.3.1 Reinforcement Learning

Figure 4.6 shows the agent-environment interaction model of reinforcement learning. At the decision epoch, the agent obtains environment state s and corresponding reward r. According to the given policy, the agent selects action a to work on the current environment, which makes state s turn into new state s'. Then at the next decision epoch, the agent and environment interact in the same way.

The agent selects the action based on the policy function which defines the behavior at a given moment. A stationary random policy is defined as $\pi: S \times A \rightarrow [0, 1]$,

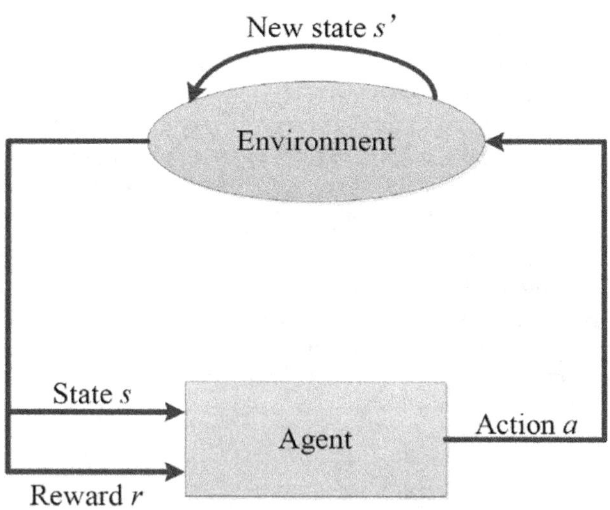

Fig. 4.6 The agent-environment interaction model of reinforcement learning

where $\pi[s, a]$ represents the probability of selecting action a under state s, S is state space, and A is action space.

There are two value functions to represent the feedbacks from each decision, namely state value function $V^{\pi}(s)$ and action-state value function $Q^{\pi}(s, a)$. $V^{\pi}(s)$ means the expected total rewards based on policy π in state s, and it can be represented as:

$$V^{\pi}(s) = E^{\pi}[\sum_{k=0}^{\infty} \gamma^k r_{t+k+1} | s_t = s], \tag{4.18}$$

where $E^{\pi}[*]$ denotes the mathematical expectation under state transferring probability $P(s, a, s')$ and policy π . r_t is immediate reward at time t. $\gamma \in (0, 1]$ denotes the discount factor to trade-off the importance of immediate reward and long-term reward.

Additionally, action-state value function $Q^{\pi}(s, a)$ represents the expected total rewards based on policy π in state-action pair (s,a), and it can be represented as:

$$Q^{\pi}(s, a) = E^{\pi}[\sum_{t=0}^{\infty} \gamma^k r_{t+k+1} | s_t = s, a_t = a]. \tag{4.19}$$

And there is a relationship between $V^{\pi}(s)$ and $Q^{\pi}(s, a)$. For a certain policy π, $V^{\pi}(s) = Q^{\pi}(s, \pi(s))$. For a stochastic policy, $V^{\pi}(s) = \sum_{a \in A} \pi(s, a) Q^{\pi}(s, a)$. Hence, $Q^{\pi}(s, a)$ can be expressed by $V^{\pi}(s)$ as follows:

$$Q^{\pi}(s, a) = R(s, a) + \gamma \sum_{s' \in S} P(s, a, s') V^{\pi}(s'), \tag{4.20}$$

where $R(s, a) = \sum_{s' \in S} P(s, a, s') R(s, a, s')$ means the expected reward of selecting action a under state s. $R(s, a, s') = E[r_{t+1} + \gamma r_{t+2} + \gamma^2 r_{t+3} + \ldots | s_t = s, a = a_t = \pi(s_t), s_{t+1} = s']$ denotes the expected reward of selecting action a to transfer the state from s to s'.

4.2.3.2 Reinforcement Learning Formulation

In order to obtain the optimal policy, it is necessary to define state space, action space and reward function in Q-learning model. Before this, we will give the optimization problem formulation in the RL based QoS-enabled load scheduling problem.

4.2.3.3 Optimization Problem Formulation

Our target is to find out the optimal scheme to allocate *Packet-in* messages from the data plane to the control plane with the minimum waiting time of QoS flows and the acceptable packet loss rate of best-effort flows. The minimization problem can be formulated as the weighted sum as follows.

$$min \quad k_1 \sum_{t=0}^{T} \sum_{i=1}^{N_1} TQ_i^1(t) + k_2 \sum_{t=0}^{T} \sum_{k=1}^{N_2} PLQ_k^2(t), \tag{4.21}$$

subject to

$$PLQ_i^1(t) = 0, \qquad \forall i = 1, 2, \ldots N_1, \tag{4.22}$$

where we assume there are T time slots during the whole system, which starts when the first *Packet-in* message comes and terminates when the last *Packet-in* message departs. Let $t \in \{0, 1, 2, \ldots, T-1\}$ denote the time instant. $TQ_i^1(t)$ is the waiting time of QoS flows i at time instant t. $PLQ_i^1(t)$ and $PLQ_k^2(t)$ are the packet loss rates of QoS flows i and best-effort flows k at time instant t, respectively. N_1 and N_2 are the total number of QoS flows and best-effort flows, respectively. k_1 and k_2 are the scale factors, and $k_1 + k_2 = 1$.

In the above optimization problem, the constraint (4.22) guarantees that the QoS flows have no packet loss.

Notably, one of the optimal target in (4.21) is to minimize the packet loss rate of best-effort flows, which is equivalently substituted by the load variation among all controllers in the remainder of this section. Because best-effort messages have the low priority, i.e., when the messages loss happens, it is probable that best-effort messages are discarded. Thus, the lower load variation leads to the lower packet loss rate of best-effort messages directly.

4.2.3.4 State Space

In order to reduce the load variation among all controllers and the waiting time of QoS flows, we propose the definition of state space as follows:

$$S = \{s = [Q_{incom}, l_c, q_c], |Q_{incom} \in Q_{level}, l_c \in L_c, q_c \in Q_c\}, \tag{4.23}$$

where

$$Q_{level} = \{1, 2\}, \tag{4.24}$$

$$L_c = \{[l_{c_1}, l_{c_2}, \cdots, l_{c_k}, \cdots, l_{c_N}]\}, \tag{4.25}$$

$$Q_c = \{[Q_{c_1}1, Q_{c_2}1, \cdots, Q_{c_k}1, \cdots, Q_{c_N}1]\}, \tag{4.26}$$

where N is the total number of controllers in the system, and c_k means the kth controller. Q_{level} means the different QoS levels of flows in this system. When $Q_{level} = 1$, this flow is QoS flow with the high-priority. When $Q_{level} = 2$, this flow is best-effort flow with the low-priority. L_c means the set of the load status of all controllers, and l_{c_k} denotes the load status of controller c_k which is calculated by (4.17) and the number of *Packet-in* messages is recorded by the info-table module. Q_c means the set of the number of QoS flows in all controllers, and Q_{c_k1} denotes the number of QoS flows in controller c_k, which is recorded by the info-table module.

4.2.3.5 Action Space

In the system, the agent has to decide how to allocate *Packet-in* message among multiple controllers. Thus, the action space A of RL can be defined as follows,

$$A = \{a_{c_1}, a_{c_2}, \cdots, a_{c_k}, \cdots, a_{c_N}\}, \tag{4.27}$$

where a_{c_k} represents the allocation control between the current *Packet-in* message and controller c_k. if $a_{c_k} = 1$, the current *Packet-in* message is assigned to controller c_k. if $a_{c_k} = 0$, the current *Packet-in* message is not assigned to the controller c_k. Note that $\sum_{k=1}^{N} a_{c_k} = 1$, which guarantees that the current *Packet-in* message only has one assigned controller.

4.2.3.6 Reward Function

We define numerical reward r that the agent obtains from taking action a at state s. We have two targets as shown in (4.21), including minimizing the load variation and the waiting time of QoS flows. Accordingly, there are two parts in reward function r, including the standard deviation of all controllers and the number of messages whose QoS levels exceed the incoming message, respectively.

The lower standard deviation means the better load balancing. Since bigger reward is taken in Q-learning, we use the negative standard deviation to represent the load variation in reward function, which is denoted in the first part of reward function r.

Since all controllers in the system are QoS-enabled, which means *Packet-in* messages will re-queue after arriving at all controllers to make sure QoS flows to be processed with the high priority. Thus, the waiting time of incoming QoS flows is only related to the number of QoS flows before them. The fewer QoS flows lead to the less waiting time, which is denoted in the second part of reward function r.

In summary, reward function r can be expressed as follows:

$$\begin{aligned} r = & - k_2 std(l_{c_1}, l_{c_2}, \cdots, l_{c_N}) \\ & - k_1 \sum_{i=1}^{N_{c_k}} bool(Q_{incom} \geq Q_{queue}(i)), \end{aligned} \tag{4.28}$$

where N_{c_k} is the number of *Packet-in* messages in controller c_k. Q_{incom} represents the QoS level of incoming message, and $Q_{queue}(i)$ means the QoS level of the ith messages in controller c_k. *bool* denotes the Boolean operator. For example, when the QoS level of the incoming message is 1 and the current messages' QoS line of one controller is 11222, the result of $\sum_{i=1}^{N_{c_k}} bool(Q_{incom} \geq Q_{queue}(i))$ is 2. In the other controller, its current messages' QoS line is 11122, the result of boolean operation is 3. If only consider the waiting time of incoming message, the agent is more possible to allocate this message to the first controller because of the less waiting time. k_1 and k_2 are the scale factors which are the same as (4.21).

4.2.3.7 Method to Learn the Optimal Strategy

Q-learning is a typical algorithm of reinforcement learning, and we use Q-learning to learn the optimal strategy in this part, where action-state value function $Q(s, a)$ is selected as the estimation function, rather than state value function $V(s)$. The basic idea of Q-learning is to evaluate $Q(s, a)$ by a temporal difference method, which is denoted as follows:

$$Q(s_t, a_t) \leftarrow Q(s_t, a_t) + \alpha(r_t + \gamma \max_a Q(s_{t+1}, a) - Q(s_t, a_t)), \tag{4.29}$$

where α denotes the learning efficiency. In Q-learning, each $Q(s, a)$ is put into Q-table.

At first, Q-learning initializes the $Q-$table. Then at state s_t, the agent determines action a_t according to ε-greedy policy, and obtains the experience knowledge as well as the training samples $(s_t, a_t, s_{t+1}, a_{t+1})$. Meanwhile, the agent uses (4.29) to update $Q(s_t, a_t)$ and $Q-$table. When meeting with the goal state, the agent terminates one loop iteration. Then Q-learning continues a new loop iteration from the initial state until the end of learning. The algorithm performed on each step is shown in Algorithm 1.

Algorithm 3 Q-learning

Initialize $Q-$table, and set parameters α, γ and k_0;
for $k = 1 : k_0$ **do**
  Select a initial state s_t randomly
  while $s_t! = s_{goal}$ **do**
    Select action a_t based on ε-greedy policy, and obtain immediate reward r_t and next state s_{t+1}
    $Q(s_t, a_t) \leftarrow Q(s_t, a_t) + (r_t + \gamma \max_a Q(s_{t+1}, a) - Q(s_t, a_t))$
    $s_t \leftarrow s_{t+1}$
  end while
end for

4.2.4 *Simulation Results and Discussions*

In this subsection, we use computer simulation to evaluate the performance of RL based QoS-enabled load scheduling. Firstly, we describe simulation settings, then present the simulation results. In this subsection, we use computer simulation to evaluate the performance of RL based QoS-enabled load scheduling. Firstly, we describe simulation settings, then present the simulation results.

4.2.4.1 Network Topology

We choose the same topology as the one in [34], which has three controllers in the control plane, and several switches in the data plane. Thus, $N = 3$.

4.2.4.2 Parameter Settings

The different seeds are employed in the simulation, and performances are average to estimate the performance of our proposed scheme. We utilize the queuing theory to model the arrival, processing and departure of *Packet-in* messages. Here, the arrival of the *Packet-in* messages is based on a Poisson distribution with the parameter of λ, indicated by the arriving rate of *Packet-in* messages. The processing time of each controller is based on the negative exponential distribution with the parameter of μ, indicated by the performance of controllers. And we assume that all controllers have the same performance, i.e., the same μ. Summarily, the values of all parameters in the simulation are summarized in Table 4.2.

Table 4.2 Parameters setting in the simulation

Parameter	Value	Description
max	20	The maximum number of *Packet-in* messages in controllers
ρ	0.5	The performance parameter of the controller
β	5	The performance parameter of the controller
θ	15	The performance parameter of the controller
N	3	The number of total controllers in the control plane
k_1	0.6	One scale factor
k_2	0.4	Another scale factor
ε	0.9	The greed factor
γ	0.65	The discount factor
Initialized α	1	The learning efficiency
μ	16	Service rate of each controller

For performance comparison, four schemes are simulated:

- RL based QoS-enabled load scheduling scheme proposed in this part and we call it as RL in the remainder of this section.
- Dynamic weight based QoS-enabled load scheduling scheme in the work of [34] and we call it as DW in the remainder of this section.
- QoS-enabled scheme, and this scheme does not take consideration of the load scheduling. We call it as QS in the remainder of this section.
- Mini-connected load scheduling scheme, and this scheme does not consider the QoS. We call it as MN in the remainder of this section.

4.2.4.3 Performance Evaluation Results

Figure 4.7 shows the relationship between the load variation and the *Packet-in* messages arriving rates of different schemes when the proportion of QoS messages is 75%. With the increase of the arriving rate, the load variation is increasing. The reason is that, as the arriving rates increasing, more messages accumulate in controllers, which obviously results in the larger load variation. In any case of arriving rate, QS's load variation is much bigger than others', because QS scheme only considers the priority of messages and fails to take the load balancing into consideration, some controllers are overloaded and others are idle, which leads to the biggest load variation. Taking the load balancing into consideration, the other three schemes' load variations are much smaller. Relatively speaking, DW's load variation is bigger. The reason is that, the adjustment of the load does not happen

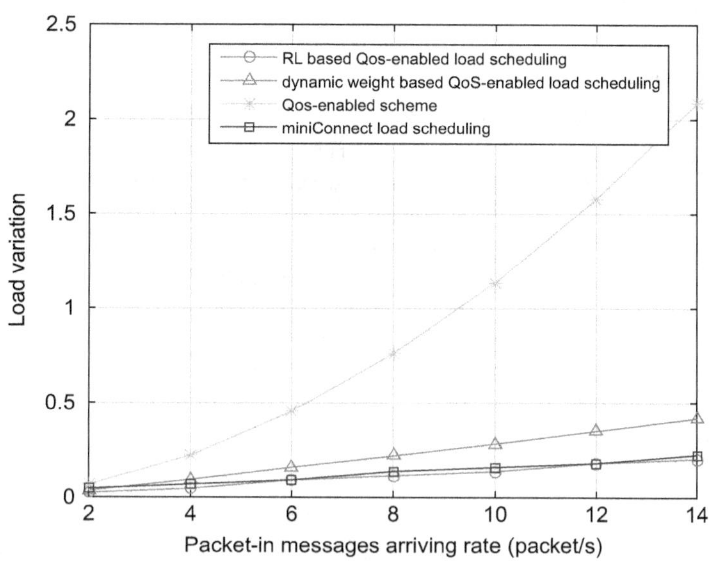

Fig. 4.7 Load variation versus *Packet-in* messages arriving rates of RL, DM, QS and MN

at each step in DW. Only when overloaded, the dynamic weight load balancing is triggered. But in MN, when each message arrives, it is assigned to the controller with the least load status, which is equivalent to adjust the load distribution scheme in each step. So MN is better than DM. But in any case of the arriving rates, RL's load variation is very close to the MN's curve. Even the RL's load variation is smaller than the MN's in some cases. Because by the offline learning of the historical data, RL performs the optimal load distribution globally, which results in the best load scheduling effect.

Figure 4.8 displays the relationship between the waiting time of QoS messages and *Packet-in* messages arriving rates of different schemes when the proportion of QoS messages is 75%. For QS scheme, although it considers the priority of messages to let the messages with the high priority be processed firstly, no load balancing mechanism also results in more waiting time of QoS messages. In the low arriving rates case, DW's waiting time is less than MN's. The reason is that, when the arriving rate is lower, it is unlikely to be overloaded in controllers, but MN scheme needs to get the load status of controllers by exchanging the signaling to adjust the load distribution in each step, which results in the additional time delay. DW scheme isn't triggered in the low load status, so it does not lead to the time delay of the signaling exchange and has relatively smaller time delay, compared with the MN scheme. With the increase of arriving rates and messages accumulating in controllers, MN and DW schemes also exchange the signaling frequently, but MN has the better load balancing performance, as shown in Fig. 4.7, it also has the better time efficiency, compared with DW. And for RL scheme, because the

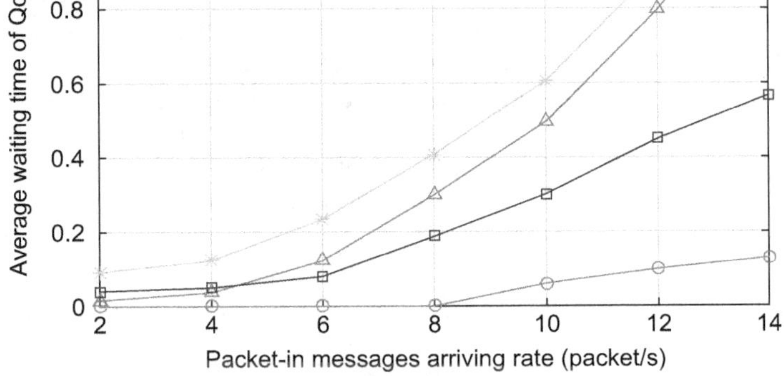

Fig. 4.8 Average waiting time of QoS messages versus *Packet-in* messages arriving rates of RL, DM, QS and MN

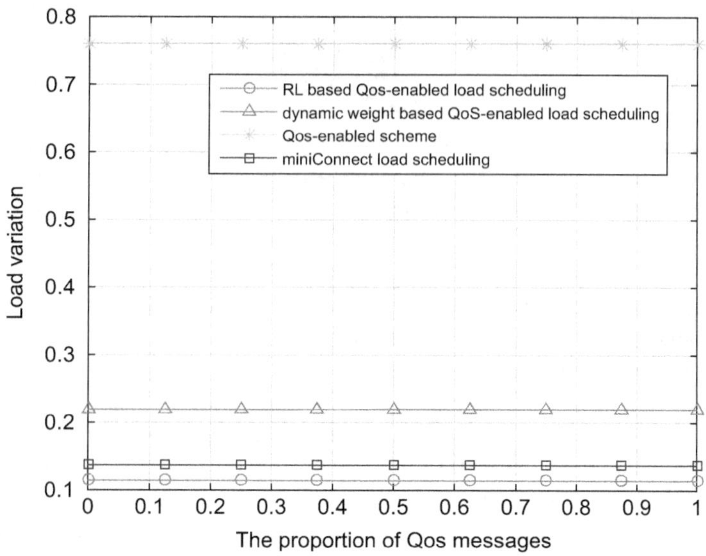

Fig. 4.9 Load variation versus the proportion of QoS messages in RL, DM, QS and MN

allocated scheme has been learned offline and in advance, and it is no need for RL scheme to exchange the signaling at all. So in the lower arriving rates, RL scheme has no additional time delay. In the higher arriving rates, RL scheme has a little time delay because of the increasing of messages. Overall, RL scheme has the best time efficiency.

Figure 4.9 presents the load variation when the proportion of QoS messages changes at the arriving rate of 8 packet/s. Because the arrival rate is constant, the load variation has no relationship with the proportion of QoS messages. But we can draw the similar conclusion as Fig. 4.7, which is that RL scheme has the best load variation.

Figure 4.10 shows the relationship between the waiting time of QoS messages and the proportion of QoS messages at the arriving rate of 8 packet/s. For QS, with the growth of QoS messages, the waiting time increases linearly, because it only considers the priority of messages but no load balancing. In the lower proportion case, DW's waiting time is less than MN's. The reason is that, only when overload happens, DW scheme is triggered. And MN scheme happens in each decision epoch. Under the current arriving rate, it is unlikely to be overloaded. So DW has relatively smaller time delay, compared with MN in the lower proportion. The increase of proportion leads to the growth of time delay directly. MN has the better load balancing as shown in Fig. 4.9, which results in the better time efficiency in the higher proportion. RL enables to learn the allocated scheme in advance and offline with no signaling exchanging. So when QoS messages are smaller, RL has no time delay completely. And with the growth of QoS messages, RL has a litter time delay and the best time efficiency.

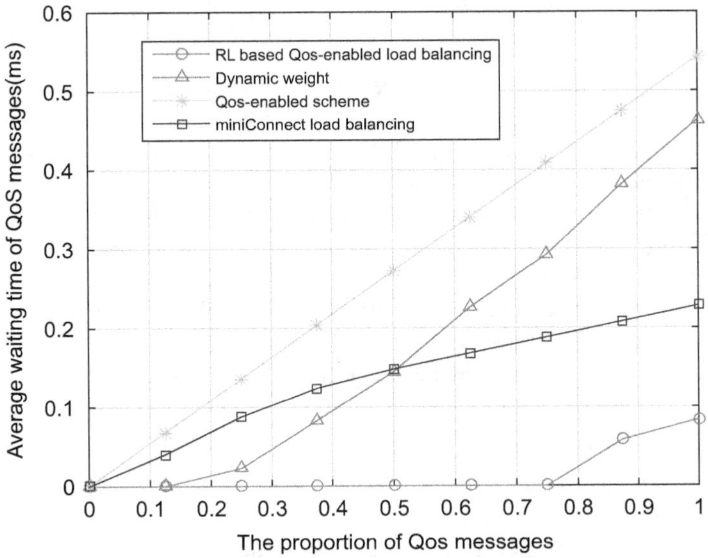

Fig. 4.10 Average waiting time of QoS messages versus the proportion of QoS messages in RL, DM, QS and MN

4.3 WLAN Interference Self-Optimization Based SOM Neural Networks

An important difference between WLAN (Wireless Local Area Networks) and other commercial mobile communication networks like 3G, 4G is that any organizations and individuals can optionally arrange their AP (Wireless Access Point) according to their own needs [66]. Due to the limitation of the spectrum and the randomness of the AP arrangement, the interference in the network has become a major issue.

Take 802.11b/g IEEE series protocol as an example. The protocol defines 13 channels which can be used by a WLAN. Each channel occupies 22 MHz bandwidth, but adjacent channel spacing is only 5 MHz. If the channel is completely orthogonal, at least five channels must be spaced between adjacent channels [67]. Then the entire spectrum can support three APs' orthogonal configurations at most. In reality, a WLAN will firstly configure itself into an optimal network environment, but in the use process, every AP channel may fluctuate. The research focus of the paper lies in how to quickly detect the AP that has a problem in self-optimization.

4.3.1 Som Network Training

In order to enable the network to have some self-organization features to reduce manual operations and maintenance [117], we use SOM to establish an artificial neural network model [26, 31, 40, 73]. We make the SINR values of all points which are simulated in space of software platform as the input vectors to study and classify, list all cases triggering interference self-optimization, and extract the vector characteristics which have trigger effects. a number of trainings are carried out. Then, the feature vectors are trained. The SOM network can cluster all characteristics of interference for self-optimization, which helps quickly determine the location of the AP fault.

The steps of applying a SOM neural network to diagnose network faults are as follows:

- Select reference fault samples.
- Learn each standard fault sample, and after learning, label the fault of the neuron with the largest output value.
- Feed samples into the SOM neural network model.
- If the output neurons in the output layer are in the same position with a standard fault sample, it indicates the sample has a corresponding fault; and if there are output neurons which are in the input layer having the location between many of the standard faults, indicating that these standards are all likely to occur, and the degree of each fault is determined by the Euclidean distance between the location of this fault and the corresponding standard fault location [74].

4.3.2 Som Neural Network Optimization

SOM Neural networks is proposed by a Finland scholar Ko-honen as a kind of Self-organizing Artificial neural network. The successful application area of SOM includes data mining, pattern recognition, signal processing and some other fields. It is based on feature clustering and high dimensional visualization [42]. SOM is generated by simulating competition, lateral inhibition which means neurons in each layer of the network can inhibit other neurons responsive to the input mode in order to make itself the winner, self-organization and other characteristics [30, 75].

Figure 4.11 shows the structure of SOM neural networks [76]. Each neuron of the input layer collects input signals from each neuron of the output layer using a weight vector. The output layer is also called a competitive layer. SOM neural network can map the input signal matrix of an arbitrary dimension to the discrete matrix of one-dimensional or two-dimensional, also can achieve this transformation by the way of topological structure [119]. Each neuron of network layer competes to corresponding opportunity, and finally, there are only one neuron in the victory [44]. Also those connection weights related to the winning neuron are adjusted toward more conducive to competition. Upon receiving a set of data, the SOM neural networks can cluster the data automatically following a specific neuron.

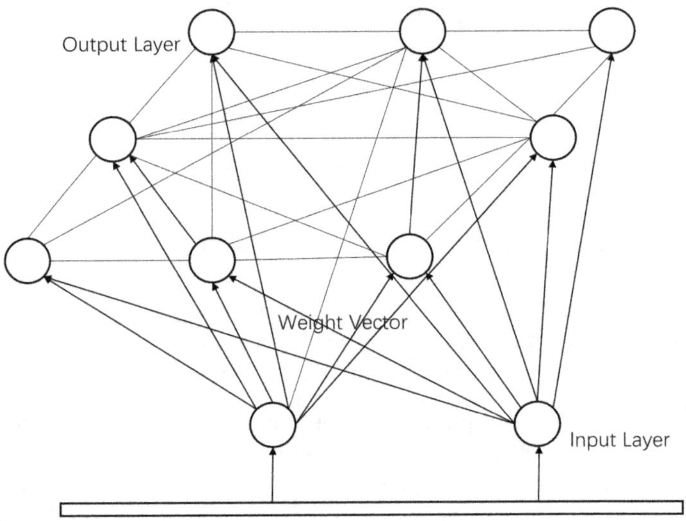

Fig. 4.11 SOM neural network

SOM clustering algorithm learning process is as follows, all the formula is based on SOM Neural Network reference [77]:

(a) Initializing network ζ_0, let n represent the dimension of the input data space, and input vector is denoted by

$$X[x_1, x_2, x_3, \ldots, x_n]^T \tag{4.30}$$

The number of synaptic weight vector equals to the dimension of input data. The weight vector of neuron J is denoted by

$$v_j = [v_{j1}, v_{j2}, v_{j3}, \ldots, v_{jn}] \qquad j \in \zeta \tag{4.31}$$

(b) Similarity matching. Search n times for the best matching (winning) neuron β by least Euclidean distance criteria.

$$\beta(X) = \arg\min |X - v_j| \qquad j \in \zeta \tag{4.32}$$

(c) Updating. Adjusting the connection between winning nodes $\beta(X)$ and theirs neighborhood nodes. Through reference [78], we can obtain the following formula:

$$v_j\,(n+1) = v_j\,(n) + \eta\,(n)\,h\beta(x), \quad j\,(n)\big(x\,(n) - v_j\,(n)\big) \tag{4.33}$$

The neighborhood function $h_{i(x),j}$ is denoted by Gaussian function $\exp(\frac{-d_{j,\beta}^2}{2\sigma^2})$ neighborhood radius $\sigma(n) = \sigma\exp(\frac{-n}{\tau_1})$ contains more adjacent nodes as far as possible, τ_1 is a time constant, σ_0 is an initial neighborhood radius; $d_{\beta,j} = |x_j - x_\beta|^2$ is the lateral distance of winning nodes β and excitatory node j.

Adjusting learning parameter $\eta(n)$ using Eq. (4.34).

$$\eta(n) = \eta_0 \exp\left(\frac{-n}{\tau_2}\right) \tag{4.34}$$

τ_2 is another time constant of SOM clustering algorithm. Starting from the initial value $\eta(0)$, $\eta(n)$ decreases gradually.

The algorithm details of the system model

First and foremost, the total model is based on the ICIC theory, connecting with the SOM neural networks. There are mainly have two cardinal line,

- One is directing at the N Aps group, do optimization at the whole group. The aim is to control the total power to be the most efficient generally.
- Another is directing to every separate AP, to control the own power of one AP. The aim is to reduce the same-frequency interference and make the power to be more efficient.

At the Beginning of the algorithm, we need to define different threshold P_i for different base station. Different P_i are corresponding to the magnitude of power for different situation of communication. For example, while considering a situation S_1 with people distributed uniformly, the amount of access and SNR and other related parameters determines a set of power thresholds P_i. Meanwhile, for a situation S_2 with dense crowds, the parameters are changed, therefore the set of power thresholds P_i is changed. Thus the basic feature is P_i is a set of value of power thresholds generated by different situation. Before doing the neural network optimization, the algorithm will ask the user to set different parameters $\alpha_1, \alpha_2, \alpha_3, \ldots$ And determine the value of P_i for different situation.

These situations may be controlled by the measurement of AP and the set of parameters $\alpha_1, \alpha_2, \alpha_3, \ldots$ But we will only discuss the situation of a specific situation and take its set of P_i value as input then focus on the historical P_i value of a single AP. We will also find different P_i as new P_i. The period of P_i update is a longtime learning period.

The learning period T cannot be too long, since it will reduce the optimization of the system a lot. Meanwhile the learning period cannot be too short either, which will waste power consumption and storage. The learning period is an important parameter. After each learning period, all the P_i will change, which means the situation is changed. Thus, T determines the working efficiecy and the optimizing efficiency [37, 46, 78].

We must emphasize that the learning period do not need to equal to switching time. Since the changing frequency of Pi may far outweigh T, P_i may be switched just-in-time (live-update) for a system that is flexible enough, once the situation is changed and be sensed by AP, the system can switch P_i. However, this is situation maybe rare in the whole SQM neural network, so it is not representative enough. In simulation, we set the switching time to be one tenth of T.

For a live-update system, AP will optimize the power consumption correspond to the threshold of P_i and achieve the minimum power consumption of single AP while still need to guarantee the quality requirement of communication.

For the Situation of N Aps group:

After each single AP study period (T) is completed, the next step is to adjust the N APs group power [48]. In this step, we must assurance the communication quality. Then, considering the N APs group as a control unit and cooperate with SDN theory [79], we use the SOM neural network algorithm to optimize two aspect of one control unit:

- Undermine the co-channel interference.
- Optimize the group power.

The first one is to optimize co-channel interference. In many circumstances, the scene outbreak near a AP will lead to the power surge. At the same time, surrounding APs also be interfered stronger by the AP. Here, the scenario could let the APs associated with the one have to raise the power [80]. However, different scenarios results in different situations. A live-update system here using SOM neural networks to optimize the co-channel interference, is the first important aspect of our software.

Secondly, it's the whole group power of the N APs domain optimization. Based on ICIC thoughts, we could consider sharing the complementary situation in the frequency domain. As an example, considering the overall power of N APs cluster region to achieve the overall optimal power. That is, in the situation of communication assurance, to control the overall power and minimum values [51, 81]. Throughout the model, we determined the frequency of center as well as edge according to the idea of ICIC model. At the same time, we make the definition of two significant factor: μ_1 and μ_2. The center-frequency power factor is μ_1, and the edge-frequency power factor is μ_2. In consideration of the overall center and edge power distribution, we do the optimization at the N APs group. We want to get the result is to achieve the optimal power distribution.

Figure 4.2 shows the SOM network training workflow more clearly. self-configuration and power self-configuration.

Process of a learning period (T):

1. Input the historical data set.
2. Input different parameter for different scenarios.
3. Use SOM network to do feature selection.
4. Apply affinity propagation clustering algorithm to the data selected, determined the P_i.
5. Update the data set, and check the communication situation.
6. Begin the N APs group co-channel interference optimizing.
7. According to the total communication situation, optimizing the total power efficiency.
8. Check every AP communication situation.
9. Predict the output and do comparison.

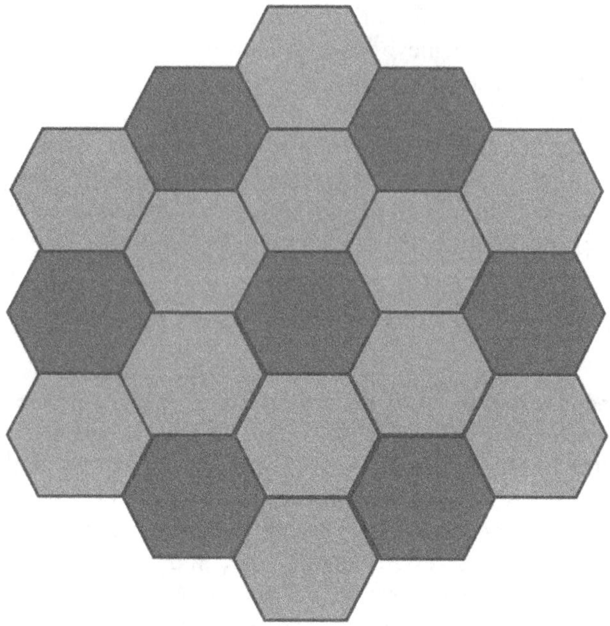

Fig. 4.12 APs' basic frequency distribution

Frequency Self-Configuration We take a more basic cellular structure to do the frequency distribution of AP. We select 1, 6, 11 channel as the AP's initial allocation channel, distinguished by colors, which shows in Fig. 4.12.

Power Self-Configuration First, we calculate the reuse distance between the same frequency, and then set the power to meet the requirements of the ratio of coverage, and finally self-configuration is divided into two conditions—the highest Energy Efficiency Ratio and the best Signal Interference Noise Ratio.

We enable the initial self-configuration of 19 APs, which is shown in Fig. 4.13:

4.3.3 Network Training And Simulation

4.3.3.1 Simulation Platform

Software Development Environment

- System: Linux (Ubuntu 14.04)
- CPU: 4 GHz
- Memory: 16 G
- Compiling environment: C++, qmake
- Packet: Qt-OpenGL

Fig. 4.13 The initial self-configuration of APs

Table 4.3 Specific functional architecture of software

Class	Function
Ap	The basic attributes of stored AP
Dot	Some basic properties of stored map points
EventButton	Events button, basic attributes of stored events
EventSet	Read-in of parameters of event editor window
GraphicsTest	Unit tests of the drawing component
Introduce	Software introduction window
Main	Start the interface
Mainwindow	Main window
Self_optimize_set	Read in the optimization of related parameters
Self_set	Algorithm call interface
Self_youhua	The realization of the self-optimization algorithm
Set_algorithm	The implementation of the self-configuration algorithm
Timer	Multithreading management of events
User	Store the basic attributes of the user

Software is constructed using "qt" which is based on C++ graphical interface application development framework, and the visual performance of AP uses the qt—OpenGL library. The specific classes are listed in Table 4.3.

System simulated to implement the related functions of self-optimization, including self-planning, self-configuration, sleep patterns, load balancing, and other issues about cover and interference.

We validate the feasibility of the theory by software simulation platform, evaluate the manifestation of our algorithm, and convert the result to the actual communication network optimization scheme to provide operators with the base station deployment, the related policy decisions.

The visualization implement of model uses the Qt—OpenGL drawing component. In prophase, we thoroughly studied and compared the OpenGL, OpenNI library and NiTE library, read the related English documents and got familiar with the basic principle of qt graphical QML programming ideas of operation, which helped to understand and call the relevant API. In view of the practical problems, our software selectively uses part of the self-configuration as well as the self-optimization to reach the purpose of judging the AP layout and allocating appropriate power.

4.3.3.2 Frequency Simulation

First, we begin with the neural network initialization settings, namely the construction of input layer neurons. We configured the 19-AP in simulation according to the ideal state of power which represents the optimum coverage without interference, and then set the SINR value of the 400 * 400 small lattice which is used as the initial reference vector of neural network learning. Feature vector training set is shown in Fig. 4.14 with the attachment lines between points representing

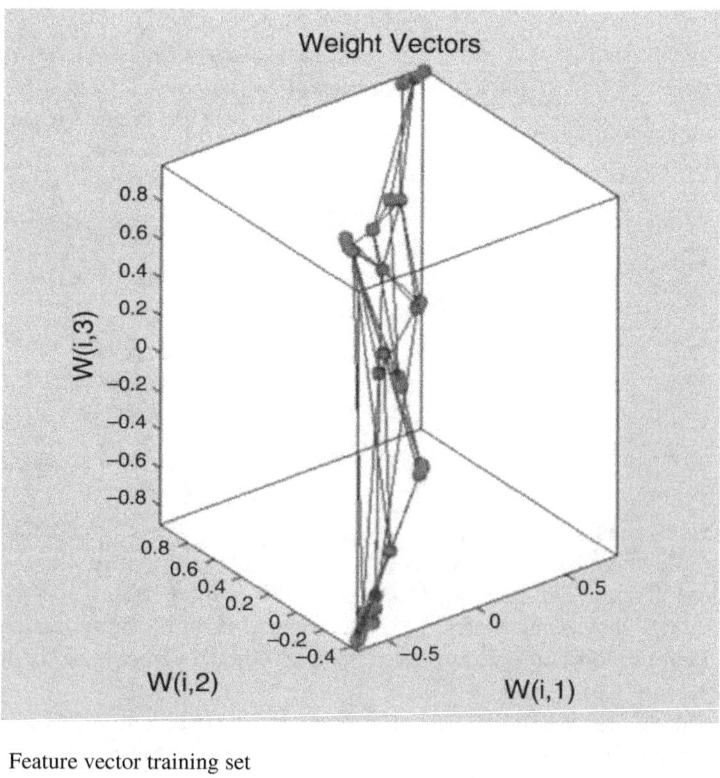

Fig. 4.14 Feature vector training set

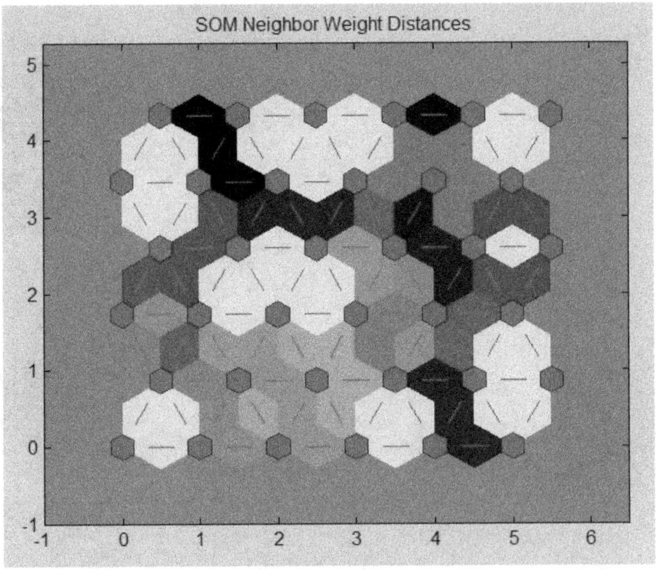

Fig. 4.15 Self-organized SOM

the resulting weight vectors. Furthermore, Fig. 4.15 is the SOM Neighborhood Weighted Distance of different positions within the region [82]. Color difference represents the discrepancy of neighborhood weighted distance value in exact area.

We plot a layer of SOM networks in Fig. 4.15, where the blue patches are expressed as neurons, and the red lines represent the direct neighbor relationships between them. We colored the neighbor blocks from black to yellow to show the association between each neuron's weight vector and its neighbors.

Then, we use the software simulation platform to randomly produce several numbers of different sets of input planes as a series of different characteristic vector, continuously input data to the network to do the mapping layer weights learning. We selected eight random and different Euler Distance (SOM Neighborhood Weighted Distance) corresponding to their input planes as shown in Fig. 4.16.

A set of subgraphs are generated. Each ith subgraph indicates the weights from the ith input to the layer's neurons. It displays black to represent zero connections, and red to represent the strongest positive connections.

We consider the influence of the situation about any one of the 19 AP's frequency changing from 1 to 11 to all the surrounding areas (400 * 400 small lattice area). Therefore, after 19 * 11 layers of iterative learning process, the SOM neural network is built. The weighted position change of the characteristic value as shown in Fig. 4.17.

In Fig. 4.17, green dots show the input vectors, and the blues are the weight vectors of each neuron. This graph is represented by red lines that connect the neighboring neurons to show how the SOM classifies the input space [83].

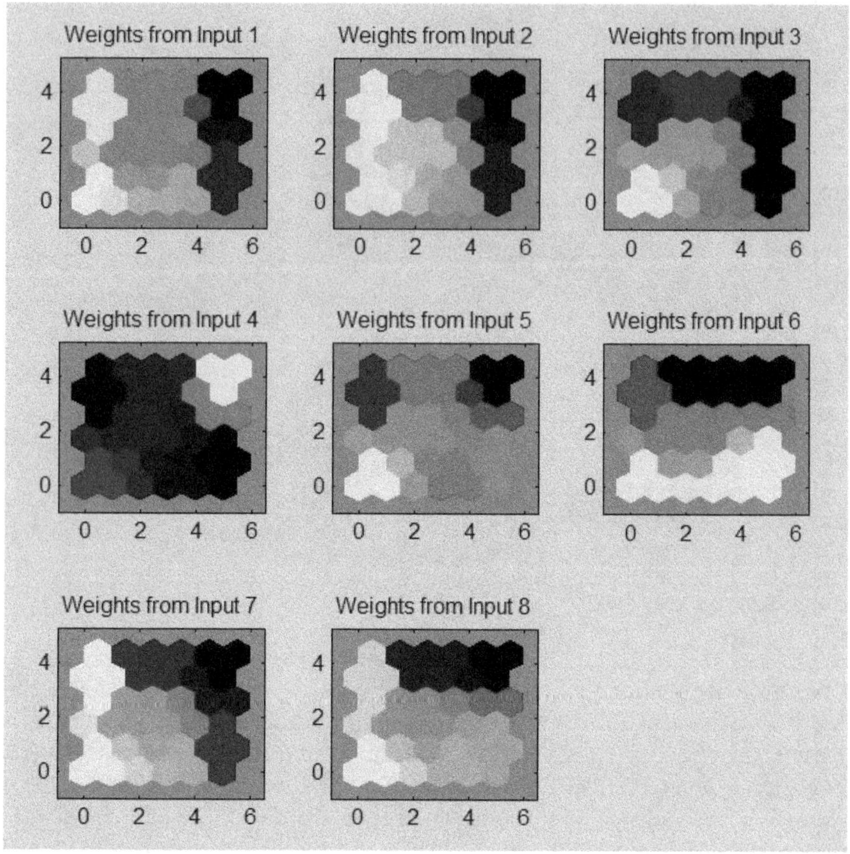

Fig. 4.16 SOM Neighbor weighted Distance according to different input planes

We further generate several multiple test sets in the simulation platform for the neural network. We keep the initial conditions and use the platform to generate a random band deployment, which is a simple hit to the wireless network. Here we can change the frequency of the AP points from 1 to 11 in the band, for example, as shown in Fig. 4.18, due to changes in the band, the software simulation platform will automatically trigger interference from optimization.

Figure 4.19 shows abnormal clustering around the SINR. It can be observed that the interference location occurs in the AP coordinates.

The results show that, in the event of co-channel interference, the SOM network can quickly find the error situation, find the AP position of interference that occurs, and determines the emergence of regional abnormal position (corresponding to the position of a small grid). Then the network provides appropriate responses to the user area [64, 84].

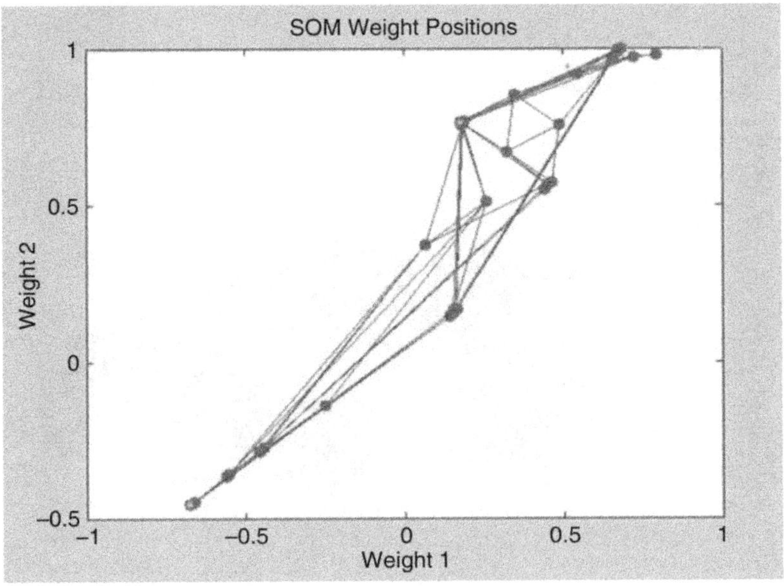

Fig. 4.17 Weighted position vector result after training

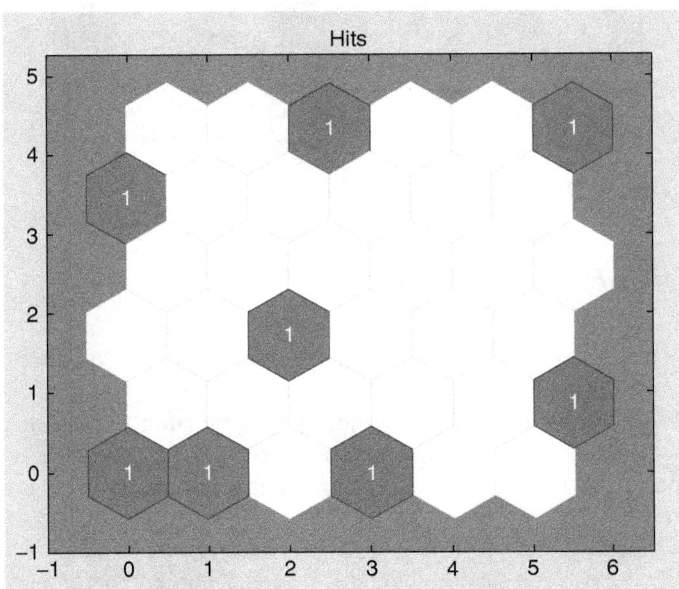

Fig. 4.18 Training result test

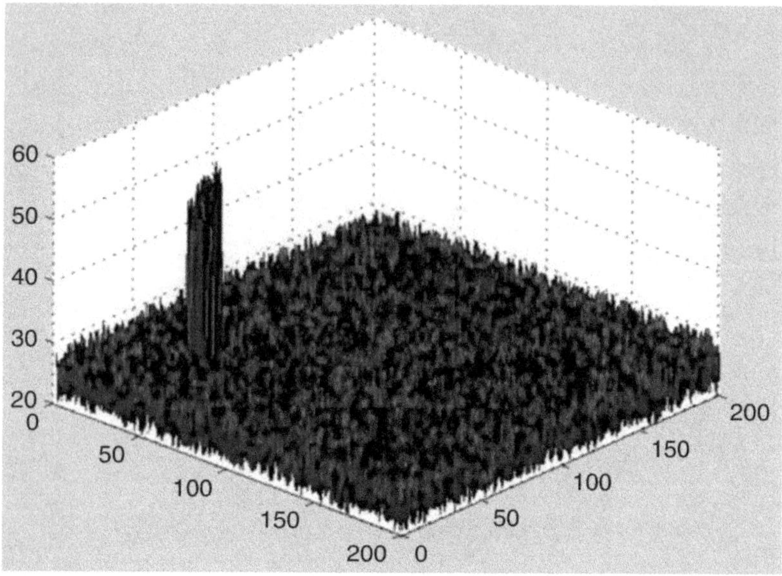

Fig. 4.19 Clustering results of SINR before and after exception

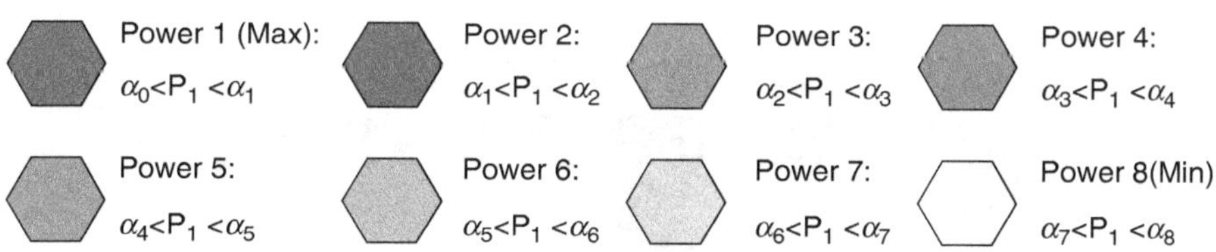

Fig. 4.20 The degree of eight power intensity symbols

4.3.4 Power Efficiency Simulation

In Fig. 4.20, here, we define eight power ranges, defined by SOM cluster situations. [85]

To power efficiency, the most important is to simulation a real situation. Here, we randomly choose six APs as the maximum power at a time which is showed in Fig. 4.21, then we do the total energy efficiency and let the network be more efficient and available.

Here, as Step 1 is showed in Fig. 4.22, we can see that in the situation of the maximum power, one AP cannot guarantee its cellar normal communication, there has to be some neighborhood APs to help its communication. It can also be observed that not every AP needs a neighborhood AP assistant, it can be considered by the real threshold-the random P_i.

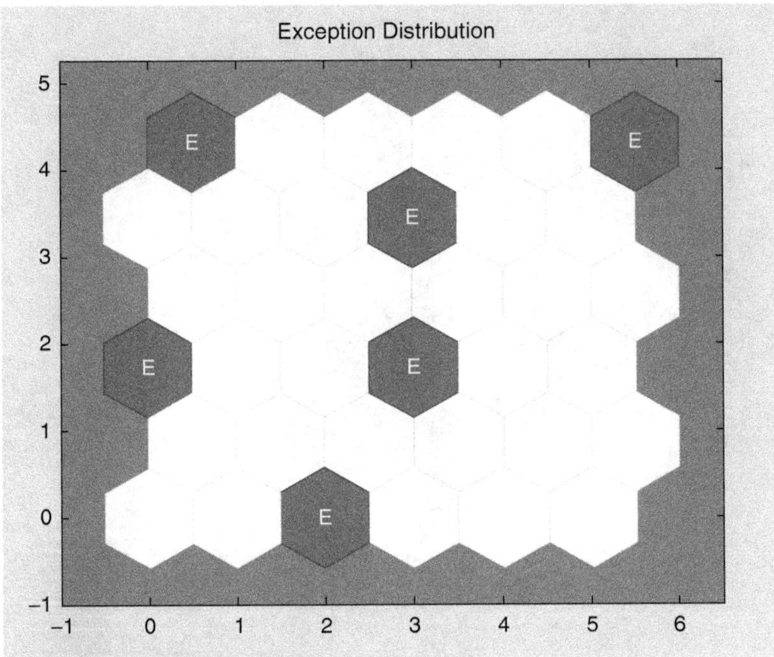

Fig. 4.21 Power efficiency of a random situation initialization

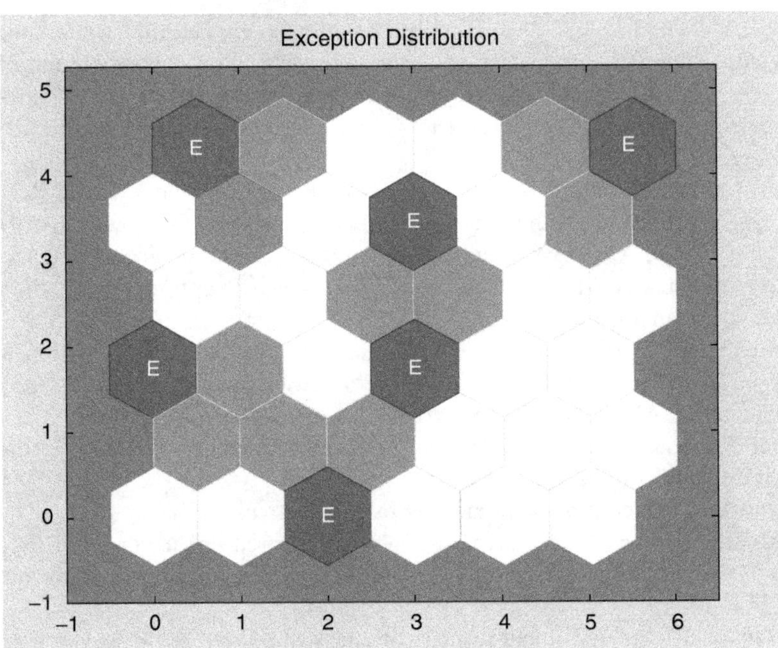

Fig. 4.22 Power efficiency of a random situation step 1

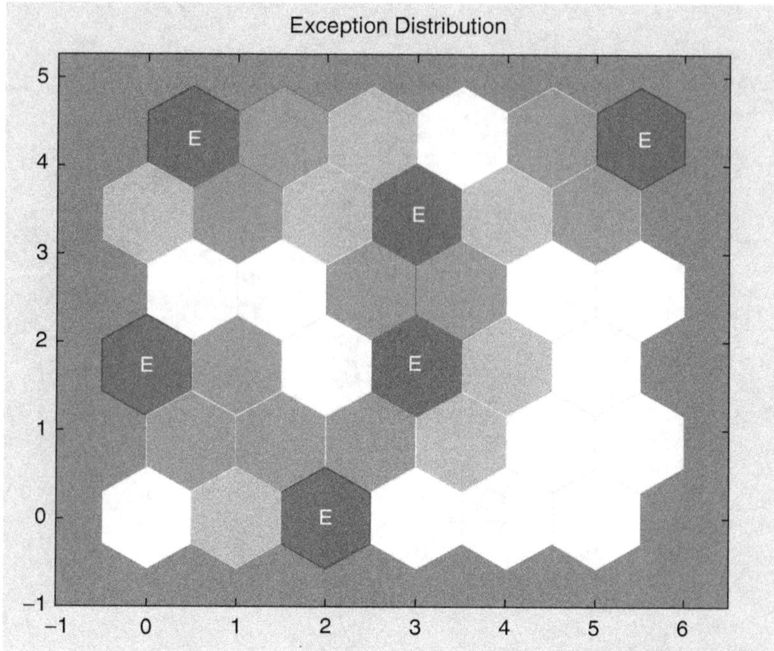

Fig. 4.23 Power efficiency of a random situation step 2

Step 2 showed in Fig. 4.23, the total network also provides the communication guarantee, in the extreme random situation it needs to do two step communication guarantee, we can see that some extreme situation also need to do second expansion and use neighborhoods to support the total environment. Here, the total P_i still keeps increasing.

Step 3 showed in Fig. 4.24, we can see that some of the extreme APs start decreasing power, and all the network begins to optimize the power distribution. In this step, all the APs have already considered the second neighborhood.

Step 4 shown in Fig. 4.25 and step 5 in Fig. 4.26 also do optimization to the whole network, and step 5 is the last step.

From the comparison with the traditional way of power distribution shown in Fig. 4.27, we can see that one AP extreme power cannot guarantee the cellar communication. Also, one AP use extreme power can lead strong interference. The different P_i, from here, we can define the standard in our software by ourselves using SOM and historical data.

Next, we have a contrast with another extreme situation.

In random situation 2, 6 extreme situations gathered in the center of Fig. 4.28, and we need to optimize the total AP. And we give the final result using SOM as shown in Fig. 4.19.

By comparing the traditional results with the result using SOM way in Fig. 4.29, we can see clearly that the SOM power efficiency is really available, both guaranteeing the communication quality, at the same time, considering the power efficiency and the total network optimization.

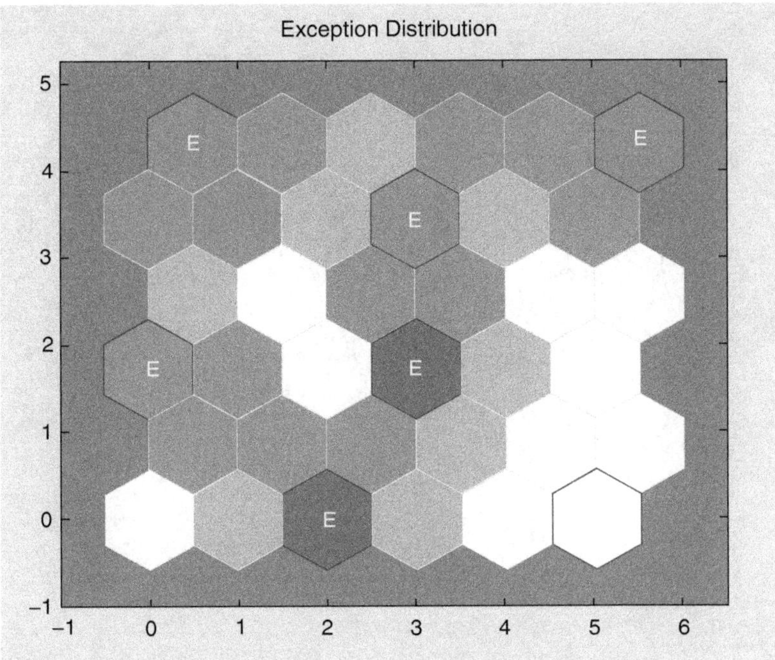

Fig. 4.24 Power efficiency of a random situation step 3

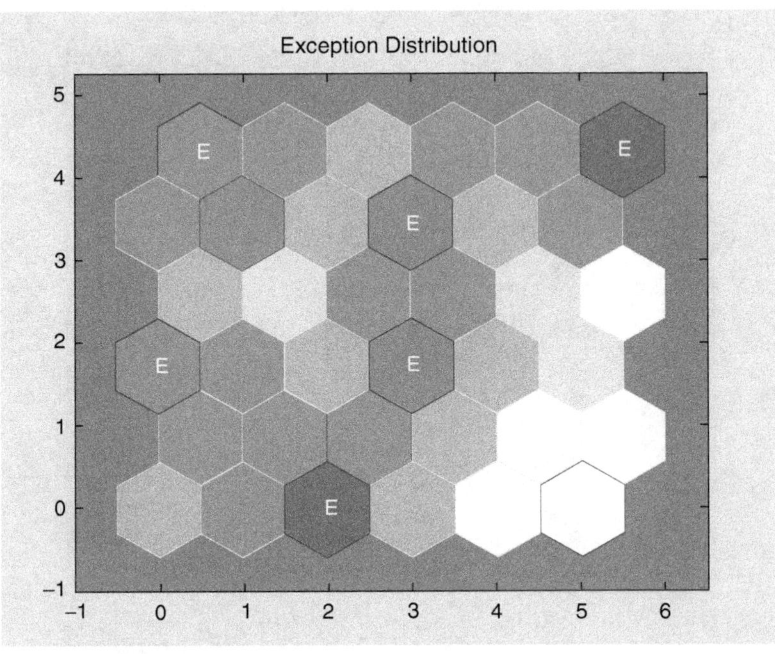

Fig. 4.25 Power efficiency of a random situation step 4

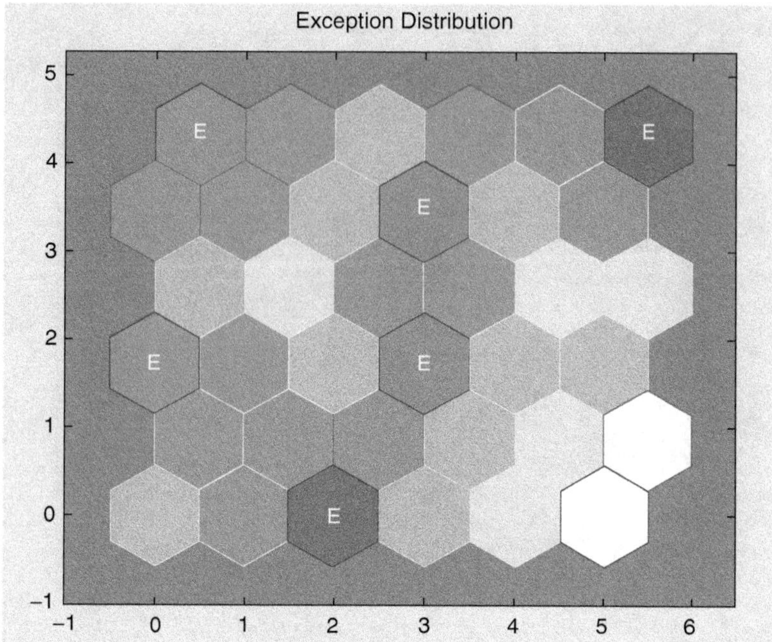

Fig. 4.26 Power efficiency of a random situation step 5

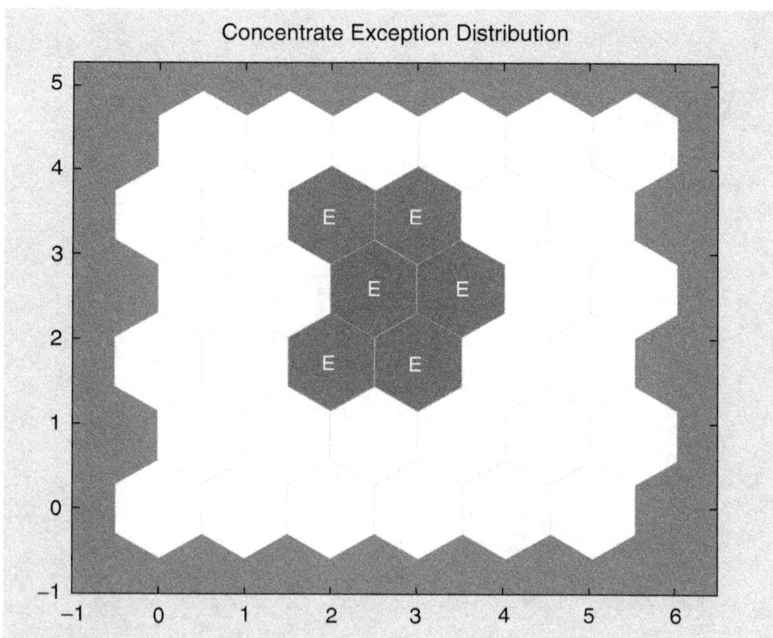

Fig. 4.27 In traditional way of power distribution

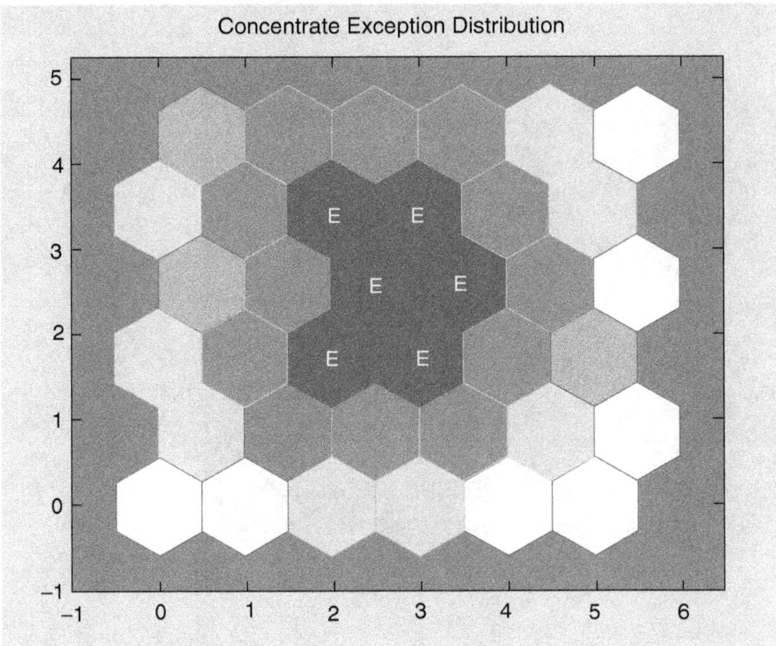

Fig. 4.28 Random situation 2 initialization

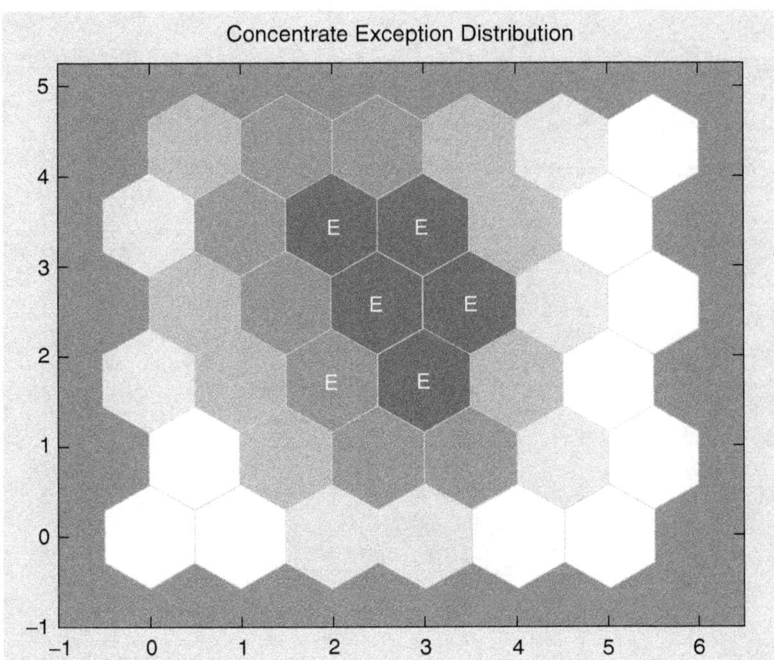

Fig. 4.29 Random situation 2 result using SOM way

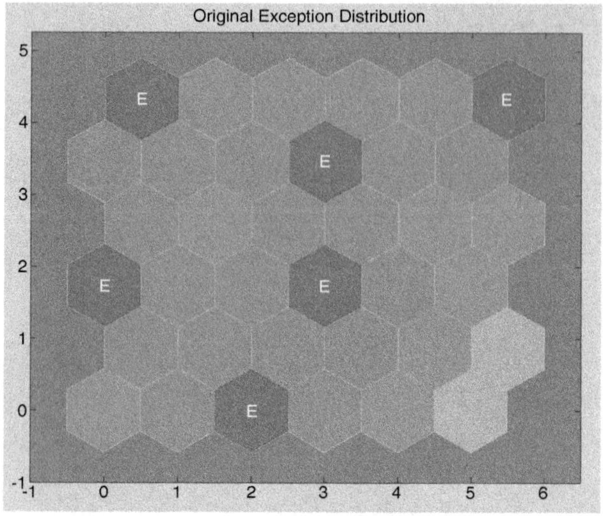

SOM	Tradition
2	6
0	0
4	21
1	0
12	7
8	2
3	0
5	0

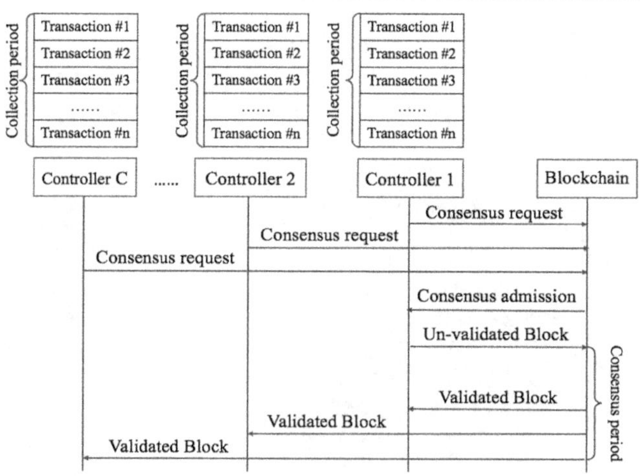

SOM	Tradition
5	6
0	0
3	12
0	0
4	1
7	2
7	8
10	7

4.4 Blockchain-Based Software-Defined Industrial Internet of Things: A Dueling Deep Q-Learning Approach

Recently, there are a growing number of applications that use Internet of Things (IoT) technologies in several industries. Industrial Internet of Things (IIoT) has emerged and attracted lots of attentions from industry and academia [91]. In order to meet the demands of high bandwidth, ubiquitous accessibility, and dynamic management, software-defined networking (SDN) [93] has been used in IIoT, called SDIIoT [94]. In addition, software-defined routing management, edge computing, flow scheduling, and energy harvesting have been researched in excellent literature [95, 97–99]. With the increasing number of industrial devices, more than one controller is employed in SDIIoT, known as distributed SDIIoT. How to reach consensus among multiple controllers is challenging in distributed SDIIoT.

Although some traditional methods enable to reach consensus among multiple controller instances, numerous non-trivial issues in the current consensus methods prevent SDIIoT from being used as a generic platform for different services and applications, including (1) the extra overheads, (2) the poor safety and liveness properties, (3) the limited scope of available network size. These challenges need to be broadly tackled by comprehensive research efforts.

Recently, *Blockchain* (BC) [35, 100] has been emerged as a novel technique, which can be used to address the above challenges. BC is a distributed ledger to record transactions, and provide trustworthy services to a group of nodes without central authority [101]. For the distributed SDIIoT, BC can act as a trusted third and 'out-of-band' party to collect and synchronize network-wide views (e.g., network events, network topology, and OpenFlow commands, etc) between different SDN controllers safely, dependably, and traceably. In general, there are two kinds of BC [103], including permissionless BC and permissioned BC. In permissionless BC, enrollment is open to anyone, and nodes enable to join and leave dynamically and frequently, using Nakamoto consensus protocol and coupled to cryptocurrency, such as proof of work (PoW) in Bitcoin [100] and proof of stake (PoS) in Ethereum [104]. All BC participants (miners) contribute their CPU power to work on an extra hard task, and only the winner of them enables to propose a block and synchronize it with others. Thus, lots of resources and time are imposed in the permissionless BC. Moreover, another BC, the permissioned BC, uses *Byzantine* fault tolerance (BFT) consensus protocol. It employs state machine replication mechanism to deal with *Byzantine* nodes that are subverted by adversaries and against the common goal of reaching consensus maliciously [105, 108], such as practical *Byzantine* fault tolerance (PBFT) [107] and Paxos [3, 109]. They always operate in a partially trusted environment, such as Hyperledger Fabric [110]. Thus, the advantages of permissioned BC are low costs, low latency, and low band-intensive. Considering partial trust, limited communication time, high loads and narrow bandwidth in SDIIoT, as well the advantages of permissioned BC, we consider to use permissioned BC in this part.

In this part, we propose a BC-based consensus protocol in distributed SDIIoT, where BC works as a trusted third party to collect and synchronize network-wide views between different SDN controllers. Specifically, it is a permissioned BC. Due to the fact that the throughput of permissioned BC is constrained with respect to the performance of BFT consensus protocol, we jointly consider the trust features of BC nodes and controllers, as well as the computational capability of BC system, so as to further improve the throughput of BC. Accordingly, we formulate view change, access selection, and computational resources allocation as a joint optimization problem. We describe this problem as a Markov decision process by defining state space, action space, and reward function. Inspired by the works to achieve the optimal decisions by improved learning methods [112–114], and the fact that it is difficult to solve this joint problem by traditional methods, we use a novel dueling deep Q-learning approach to solve this problem. A preliminary version of our work appeared in [115]. The distinct contributions of this part are as follows.

- We propose a BC-based consensus protocol to simplify and secure the collection and synchronization of network views between different SDN controllers. And we give consensus steps in detail, along with theoretical analysis.
- Due to the fact that lots of network views need to be collected and synchronized by BC system, we jointly consider the trust features of BC nodes and controllers, as well as computational capability of the BC system, so as to improve the throughput. Accordingly, we formulate view change, access selection, and computational resources allocation as a joint optimization problem. We describe it as a Markov decision process by defining state space, action space, and reward function.
- In order to address this joint problem, we propose a novel dueling deep Q-learning approach to learn the optimal strategy. Simulation results with different system parameters are presented to show the effectiveness of the proposed scheme.

4.4.1 System Model

In this subsection, we introduce the system model that we use. We first present the network model, followed by the trust feature model and the computation model.

4.4.2 Network Model

We assume that there are C controllers in distributed SDIIoT, which are represented by $\mathscr{C} = \{1, \ldots, C\}$. Each of them can communicate with the third-party BC system. This BC system consists of N nodes, i.e., physical machines, denoted by $\mathscr{N} = \{1, \ldots, N\}$. Like other robust BFT protocols [120], these N nodes are under

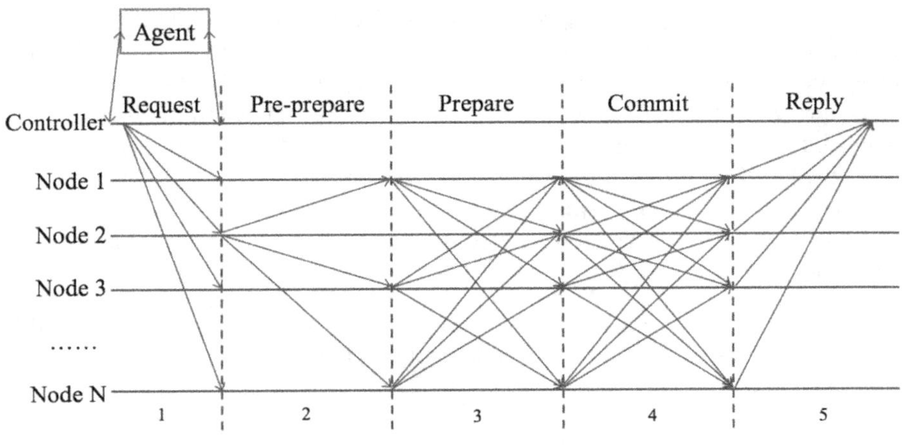

Fig. 4.30 The different network structures between traditional scheme and BC-based scheme

Byzantine failure model to make consensus, where at most f nodes are faulty [107], where $f = \frac{N-1}{3}$. And any finite number of controllers can behave arbitrarily to issue correct or incorrect transactions to the BC system. Some strong adversaries can collude with each other to compromise the replicated service. However, they can't break cryptographic technologies, i.e., signatures, message authentication code (MAC), and collision-resistant hashing. We denote the messages with cryptographic technologies as follows [38, 121].

- $\langle m \rangle_{\sigma_i}$ means that message m is signed with a pubilc key from node i.
- $\langle m \rangle_{\sigma_{i,j}}$ means that message m is authenticated by node i with a MAC for node j.
- $\langle m \rangle_{\sigma_{\mathrm{i}}}$ means that message m is authenticated by an array of MACs with node i for every replicas.

Figure 4.30 shows the different network structures between traditional scheme and BC-based scheme. It is worth mentioning that we use edge computing servers to do some computations related to the above cryptography so as to improve the throughput of BC system. There are E edge computing servers, and the set of computing servers is represented by $\mathscr{E} = \{1, 2, \ldots, E\}$.

4.4.2.1 Trust Feature Model

We consider trust features of nodes and controllers in the system. Due to the lack of centralized security services and prior security association, all nodes and controllers have diverse trust features, such as safe or compromised. It is barely possible to exactly know what the trust feature is for one node or one controller at the next time instant. Thus, the trust features of a node $n \in \{1, 2, \ldots, N\}$ and a controller $c \in \{1, 2, \ldots, C\}$ can be modelled as random variables δ^n and η^c. δ^n and η^c can be divided into discrete levels, denoted by $\xi = \{\xi_0, \xi_1, \ldots, \xi_{L-1}\}$, and

$\mathscr{D} = \{\mathscr{D}_0, \mathscr{D}_1, \ldots, \mathscr{D}_{H-1}\}$, respectively, where L and H are the number of available trust features for nodes and controllers. We assume the trust features realization of δ^n and η^c to be $\delta^n(t)$ and $\eta^c(t)$ at time slot t, respectively. There are T time slots during the period of time, which starts from when the controller issues an unvalidated block, and terminates when the controller is replied with a validated block. Let $t \in \{0, 1, 2, \ldots, T-1\}$ denote the time instant.

Considering the time correlation of real trust features in BC nodes and controllers, we use Markov chain to model the transition of trust features in BC nodes and controllers, as follows:

- For node n, let the transition probability of $\delta^n(t)$ from one state $\mathscr{X}_s$ to another state $\mathscr{Y}_s$ be $\kappa_{\mathscr{X}_s\mathscr{Y}_s}(t)$. The $L \times L$ transition probability matrix $\mathscr{K}^n(t)$ of the trust feature in node n is denoted as

$$\mathscr{K}^n(t) = [\kappa_{\mathscr{X}_s\mathscr{Y}_s}(t)]_{L\times L}, \tag{4.35}$$

  where $\kappa_{\mathscr{X}_s\mathscr{Y}_s}(t) = Pr(\delta^n(t+1) = \mathscr{Y}_s|\delta^n(t) = \mathscr{X}_s)$, and $\mathscr{X}_s, \mathscr{Y}_s \in \xi$.
- For controller c, the transition probability of $\eta^c(t)$ is $\gamma_{\theta_s\phi_s}(t)$. The $H \times H$ transition probability matrix $\Upsilon^c(t)$ of the trust feature in controller c is denoted as

$$\Upsilon^c(t) = [\gamma_{\theta_s\phi_s}(t)]_{H\times H}, \tag{4.36}$$

  where $\gamma_{\theta_s\phi_s}(t) = Pr(\eta^c(t+1) = \phi_s|\eta^c(t) = \theta_s)$, and $\theta_s, \phi_s \in \mathscr{D}$.

4.4.2.2 Computation Model

There are a number of computation tasks in the BC nodes, such as verifying signatures, generating MACs, and verifying MACs. Let $T_m = \{s_m, q_m\}$ denote a computation task related to message m, where s_m means the size of message m, and q_m is the required number of CPU cycles to complete this task.

In order to improve the throughput of BC, we use virtual computing resources from edge computing servers to do computation tasks in the BC system. There are lots of edge computing servers and some other computation tasks that also use the computation resources in edge computing servers, thus we don't exactly know the computational resources for the BC system at the next time instant. Therefore, we model the computation resources of edge computing server e for the BC system as a random variable ζ^e. To discretize the values of computation capabilities, ζ^e can be partitioned into Y discrete intervals as $\mathscr{Y} = \{\mathscr{Y}_0, \mathscr{Y}_1, \ldots, \mathscr{Y}_{Y-1}\}$. The computation resources of edge computing server e at time slot t can be denoted as $\zeta^e(t)$, $t \in \{0, 1, 2, \ldots, T-1\}$. Considering the time correlation of computation state in edge computing server, we use a Markov chain to model the transition of computation state in edge computing server. Based on a certain transition probability, $\zeta^e(t)$

changes from one state to another. Let $\vartheta_{a_s b_s}(t)$ denote the transition probability. The $Y \times Y$ computation state transition probability matrix is represented as:

$$\Pi^n(t) = [\vartheta_{a_s b_s}(t)]_{Y \times Y}, \tag{4.37}$$

where $\vartheta_{a_s b_s}(t) = Pr(\zeta^e(t+1) = b_s | \zeta^e(t) = a_s)$, and $a_s, b_s \in \mathscr{Y}$.

The execution time of computation task T_m can be denoted as

$$t_m = \frac{q_m}{\zeta^e(t)}. \tag{4.38}$$

Thus, the computation rate is

$$CompR^e(t) = a^e(t)\frac{s_m}{t_m} = a^e(t)\frac{\zeta^e(t)s_m}{q_m}, \tag{4.39}$$

where $a^e(t)$ means whether or not edge computing server e is allocated to the BC system at time slot t. $a^e(t) = 1$ denotes edge computing server e is allocated to the BC system; otherwise $a^e(t) = 0$. At one time slot, there is only one edge computing server allocated to the BC system, thus $\sum_{e=1}^{E} a^e(t) = 1$.

4.4.3 Blockchain-Based Consensus Protocol

We have presented that the existing consensus protocols are challenging in SDIIoT, and BC could be a potential approach to address these issues in the previous subsection. In this section, we propose a novel BC-based consensus protocol in distributed SDIIoT. We begin with an overview of BC-based consensus protocol. Then, we present the detailed steps of the consensus protocol, along with theoretical analysis.

4.4.3.1 Overview of BC-Based Consensus Protocol

Each controller collects its local events and OpenFlow commands as Transaction #1, Transaction #2, ..., Transaction #n, which is called the collection period. The format of a transaction is shown in Table 4.4. The number of this transaction denotes the position of this transaction. The signature and MAC make sure the integrity and authentication of this transaction. The payloads include local events and OpenFlow commands that need to be synchronized between different SDN controllers.

After the collection period, all controllers issue consensus requests to the third-party BC system. According to a policy, which will be introduced in the following subsection, called access selection, the BC system only enables one controller to access, and replies it by an admission message [39]. Then, this controller sends an

Table 4.4 The format of a transaction

The number of this transaction in the block
The Signature of this transaction
The MAC of this transaction
Payloads, including local events and OpenFlow commands

Table 4.5 The format of a block

Field	Description
Version	Block version number
Timestamp	Creation time of this block
Controller ID	The identifier of this controller
Block ID	The identifier of this block
Block payload	Transactions in this block (Transaction #1, . . . , Transaction #n)

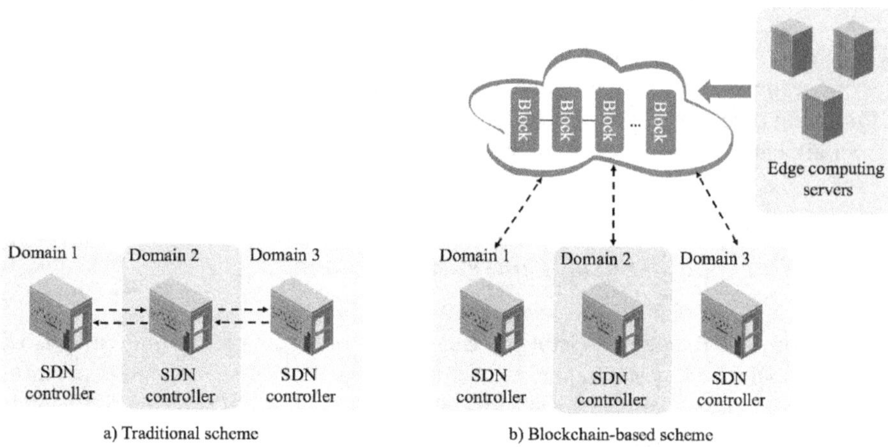

Fig. 4.31 The overview of consensus procedures in BC-based consensus protocol between different SDN controllers

un-validated block with block header and transactions, whose format is presented in Table 4.5. After reaching consensus, the BC system sends the corresponding validated block to the entire controllers. Finally, all controllers learn the payloads in each transaction to know the events and OpenFlow commands from other controller. These steps are in the consensus period. By this way, network-wide views can be synchronized between different SDN controllers.

For a comprehensive perspective, Fig. 4.31 offers the consensus procedures in blockchain structure. Here, controller 1 is the selected controller to access to the BC system.

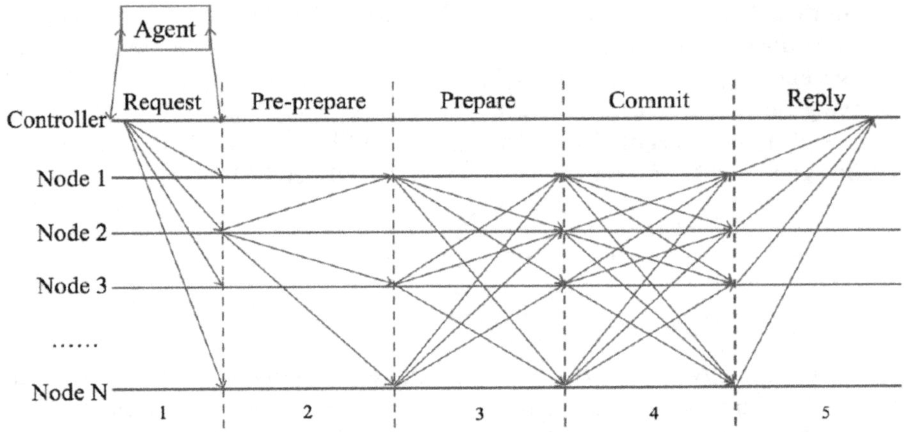

Fig. 4.32 The detailed procedures inside the permissioned BC

4.4.4 *Detailed Steps and Theoretical Analysis*

After giving the overview of BC-based consensus protocol, we will introduce the detailed steps inside the permissioned BC, along with the theoretical analysis of each step.

Based on PBFT [107], the detailed steps inside the permissioned BC is depicted in Fig. 4.32. The numbering of each step in this figure is the same as the one used in the remainder of this subsection. PBFT protocol has been used in real scenario, such as Hyperledger Fabric project [41, 110], and Hyperledger Indy project [122]. They are hosted by The Linux Foundation, and develop applications with a modular architecture. Thus, we consider that it can be used in real scenario when meeting with SDN controllers.

1. The Controller Sends an Un-Validated Block to all Nodes The selected controller sends an un-validated block to the BC system. The agent chooses one node in the BC as the primary node p. Making decisions about which node is the primary node is known as view change protocol. The view change protocol used in our proposed scheme will be introduced in the following subsection. This selected controller sends a block message $\langle\langle block\rangle_{\sigma_c}, c\rangle_{\sigma_c}$ to all nodes, where c denotes the controller ID. It is encrypted with the private signature of controller c, and authenticated with MACs for all nodes. When receiving this message, only primary node p verifies the MAC. If valid, the signature will be verified then. If still valid, it verifies the signature and the MAC of each transaction in this block, then moves to the following steps. The success rate of all verification will be recorded by the agent. If this block has already been executed, the primary will resend the validated block to this controller.

Theoretical Analysis We consider an uncivil execution during which a fraction g of transactions sent by the controller are correct [121]. The more trusted controller has the bigger g. We assume verifying one signature, generating one MAC, and verifying one MAC require θ, α, and α cycles, respectively, and the controller issues batches of transactions of size b. As the work in [43, 107], we ignore the cost of sending and receiving transactions. Therefore, the cost at the primary is

$$(1 + \frac{b}{g})(\theta + \alpha), \tag{4.40}$$

and there is no cost at non-primary nodes.

2. The Primary Node Multicasts PRE-PREPARE Message to Other Replica Nodes Finishing the verification, primary p sends a PRE-PREPARE message to all other replica nodes, as $\langle PRE{-}PREPARE, p, c, H(m)\rangle_{\sigma_\mathbf{p}}$, which is authenticated with MACs for each replica node. Here, p and $H(m)$ mean the primary node ID and the hashed result of the issued block, respectively. The replica node verifies the MAC of primary p, as well as the signature and the MAC of each transaction. Then it enters the following steps.

Theoretical Analysis In this phase, primary p generates $(N - 1)$ MACs for all replicas. Each replica verifies one MAC from primary p, as well as $\frac{b}{g}$ signatures and MACs from transactions. Thus, the cost at the primary is

$$(N - 1)\alpha, \tag{4.41}$$

and the cost at each replica node is

$$\alpha + \frac{b}{g}(\theta + \alpha). \tag{4.42}$$

3. The Replicas Send PREPARE Message to Others After verifying the validity of MACs and signatures, each replica replies the PRE-PREPARE message with sending a PREPARE message to all nodes, as $\langle PREPARE, p, c, H(m), n\rangle_{\sigma_n}$, where n denotes the replica node ID. When each replica node collects $2f$ matching PREPARE messages with its local PRE-PREPARE message, it will enter the following steps.

Theoretical Analysis In this phase, primary p needs to verify $2f$ MACs. Each replica node generates $(N - 1)$ MACs and verifies $2f$ MACs. Therefore, the cost at the primary is

$$2f\alpha, \tag{4.43}$$

and the cost at each replica is

$$(N - 1 + 2f)\alpha. \tag{4.44}$$

4. All Nodes Send COMMIT Message to Others Following the reception of $2f$ matching PREPARE messages, node n sends a COMMIT message to all others, as $\langle COMMIT, p, c, H(m), n\rangle_{\sigma_{\mathbf{n}}}$. After receiving $2f$ matching COMMIT messages, it will enter the following steps.

Theoretical Analysis In this phase, primary p needs to generate $(N-1)$ MACs, and verify $2f$ MACs. Each replica generates $(N-1)$ MACs, and verifies $2f$ MAC. Therefore, the costs at the primary and the replica are both

$$(N-1+2f)\alpha. \tag{4.45}$$

5. The Nodes Send the Validated Block to all Controllers Node n sends a REPLY message $\langle REPLY, block, n\rangle_{\sigma_{n,c}}$ to all controllers, where $block$ is the validated block. When each controller receives $2f$ valid and matching REPLY messages, it accepts this validated block and updates the corresponding network views.

Theoretical Analysis In this phase, the primary and the replicas need to generate $\frac{b}{g}$ MACs for one controllers. Therefore, the total costs at the primary and the replica are both

$$\frac{b}{g}C\alpha. \tag{4.46}$$

Thus, for one transaction, the cost at the primary is

$$(\frac{1}{g}+\frac{1}{b})\theta+(\frac{1}{b}+\frac{C+1}{g})\alpha+\frac{2N+4f-2}{b}\alpha \tag{4.47}$$

For one transaction, the cost at the replica is

$$\frac{1}{g}\theta+\frac{2+C}{g}\alpha+\frac{2N+4f-2}{b}\alpha \tag{4.48}$$

We assume that multi-core computation modules run in parallel on distinct cores. Each core has the computation speed of φ Hz. In addition, we consider an uncivil execution where the primary is not fully trusted. The primary has k trust to affect the system performance, and $k \in [0, 1]$. The more trusted primary node has the bigger k. Therefore, the throughput of the consensus protocol is at most

$$\min[\frac{k\varphi}{(\frac{1}{g}+\frac{1}{b})\theta+(\frac{1}{b}+\frac{C+1}{g})\alpha+\frac{2N+4f-2}{b}\alpha}, \frac{k\varphi}{\frac{1}{g}\theta+\frac{2+C}{g}\alpha+\frac{2N+4f-2}{b}\alpha}]trx/s. \tag{4.49}$$

4.4.5 *Problem Formulation*

We have presented the throughput of the BC system. In order to improve the throughput, we need to make the joint decisions about view changes, access selection, and computational resources allocation. In this subsection, we formulate this joint problem as a Markov decision process by defining state space, action space, and reward function.

4.4.5.1 State Space

In order to improve the throughput, the learning agent needs to sense state $s(t)$ at time slot t. As we have mentioned, the learning agent should make the joint decisions about view changes, access selection, and computational resources allocation. Accordingly, the learning agent needs to sense the trust features of all nodes and controllers, as well as the computational capabilities of all edge computing servers. Therefore, the state space can be represented as follows.

$$s(t) = \begin{bmatrix} \delta^1(t)\ \delta^2(t) \ldots \delta^N(t) \\ \eta^1(t)\ \eta^2(t) \ldots \eta^C(t) \\ \zeta^1(t)\ \zeta^2(t) \ldots \zeta^E(t) \end{bmatrix}. \tag{4.50}$$

4.4.5.2 Action Space

The agent mainly needs to decide view changes (i.e., which node is the primary node), access selection (i.e., which controller can access to the BC system), and computational resources allocation (i.e., which edge computing server should be allocated to the BC system). Thus, the action space is denoted by

$$A(t) = \{A^N(t), A^C(t), A^E(t)\}, \tag{4.51}$$

where $A^N(t)$, $A^C(t)$, and $A^E(t)$ represent:

- $A^N(t) = [a^1(t), a^2(t), \ldots, a^n(t), \ldots, a^N(t)]$, which means whether or not node n is the primary. And $a^n(t) \in \{0, 1\}$, where $a^n(t) = 1$ means node n is the primary node, otherwise it is the replica node. Note that the BC system only has one primary node, thus $\sum_{n=1}^{N} a^n(t) = 1$.
- $A^C(t) = [a^1(t), a^2(t), \ldots, a^c(t), \ldots, a^C(t)]$ decides which controller can access to the BC system. And $a^c(t) \in \{0, 1\}$, where $a^c(t) = 1$ represents controller c can access, otherwise $a^c(t) = 0$. Note that at one time slot, only one controller enables to access to the BC system, thus $\sum_{c=1}^{C} a^c(t) = 1$.
- $A^E(t) = [a^1(t), a^2(t), \ldots, a^e(t), \ldots, a^E(t)]$ determines which edge computing server is allocated to the BC system. And $a^c(t) \in \{0, 1\}$, where $a^e(t) = 1$ denotes edge computing server e is allocated, otherwise $a^e(t) = 0$. Similarly, $\sum_{e=1}^{E} a^e(t) = 1$.

4.4.5.3 Reward Function

In order to improve the throughput, we model the throughput of the BC system as reward function. According to (4.49), state space, and action space, we define the reward function as:

$$r(t) = \min[\frac{k'\varphi'}{(\frac{1}{g'} + \frac{1}{b})\theta + (\frac{1}{b} + \frac{C+1}{g'})\alpha + \frac{2N+4f-2}{b}\alpha}, \frac{k\varphi}{\frac{1}{g'}\theta + \frac{2+C}{g'}\alpha + \frac{2N+4f-2}{b}\alpha}]trx/s, \tag{4.52}$$

where $k' = \sum_{n=1}^{N} a^n(t)\delta^n(t)$, $\varphi' = \sum_{e=1}^{E} a^e(t)CompR^e(t)$, and $g' = \sum_{c=1}^{C} a^c(t)\eta^c(t)$.

Based on the above problem formulation, the learning agent senses state $s(t)$ at time slot t, and outputs a policy π that determines which action $a(t)$ should be taken. Then this action will be executed, i.e., one controller will enable to access to the BC system, one node will be the primary node, and one edge computing server will be allocated to the BC system. In order to let the learning agent remember the experience and act better next time, the immediate reward $r(t)$ will be fed back to the learning agent. The trust features of nodes and controllers, as well as the computational capabilities of edge computing servers change to next state $s(t+1)$. Then the learning agent senses them and outputs another new policy, and so on so forth [47]. The final goal is to achieve the maximum long-term reward. Summarily, the interaction of the learning agent and the environment is shown in Fig. 4.33.

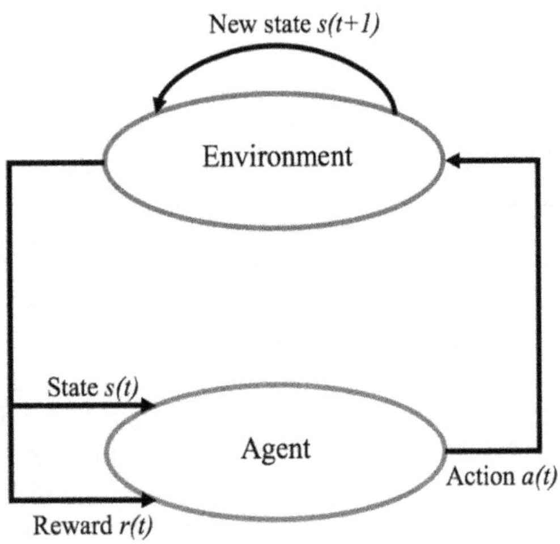

Fig. 4.33 The interaction of the learning agent and the environment

There are some challenges to solve the above problem formulation as follows.

1. The target is to maximize the long-term reward by step-and-step control. However, the learning agent only senses the state at time slot t, and the action taken at time slot t will affect the environment at time slot $t + 1$. The state cannot be obtained ahead of time. Thus, the traditional optimization method, only considering current state, is not feasible.
2. Considering the trust features of nodes and controllers, as well as the computational capabilities of edge computing servers, the system is high-dimensional and high-dynamical. It is hard to make the joint and optimal decisions by traditional methods.
3. In the BC system, taking which action has no relationship with what happens in the next time slot. For example, when the learning agent selects one controller to access to the BC system, its trust feature in the next time slot still changes according to its transition probability matrix, not the action [49]. Therefore, the traditional optimization method, learning the relationship between states and actions, is not suitable.

To address the above challenges, we will propose a dueling deep Q-learning approach in the next subsection to achieve the maximum long-term reward.

4.4.6 *Dueling Deep Q-Learning*

After the problem formulation, we consider a dueling deep Q-learning approach to address this problem. In this subsection, we begin with the introduction of Q-learning and deep Q-learning. Then we present dueling deep Q-learning that is used in this section.

4.4.6.1 Q-Learning

In the Q-learning model, the agent interacts with the environment by perceptions and actions. In one interaction step, the agent receives current state $s(t)$ from the environment, then selects an action $a(t)$ as the output, and the value of this action is measured by a scalar reward $r(t)$. This action generates next state $s(t+1)$. The agent selects actions to obtain the maximum long-term rewards. It learns to do this over several interaction steps by systematic trials and errors, guided by Q-learning [123]. Q-learning is a model-free algorithm using delay rewards. It aims to find a policy π, mapping states and actions, to maximize the long-term rewards.

There are two popular approaches to denote the feedback from each step in terms of long-term rewards, namely state-value function $V^{\pi}(s)$ and action-state value function $Q^{\pi}(s, a)$. $V^{\pi}(s)$ means the expected total reward in state s:

$$V^{\pi}(s) = E^{\pi}[\sum_{k=1}^{\infty} \gamma^k r_{t+k+1} | s_t = s], \tag{4.53}$$

where $E^{\pi}[*]$ means mathematical expectation, r_{t+k+1} means the immediate reward at time slot $t+k+1$, and $\gamma \in (0, 1)$ is discount factor to balance immediate reward and future reward. Moreover, $Q^{\pi}(s, a)$ denotes the expected total rewards in state s and action a:

$$Q^{\pi}(s, a) = E^{\pi}[\sum_{k=1}^{\infty} \gamma^k r_{t+k+1} | s_t = s, a_t = a]. \tag{4.54}$$

Q-learning evaluates $Q(s, a)$ by a temporal difference method as follows.

$$Q(s, a) \leftarrow Q(s, a) + \alpha(r + \gamma \max_{a'} Q(s', a') - Q(s, a)), \tag{4.55}$$

where α is learning rate, and $\alpha \in (0, 1]$. The action with maximum $Q(s, a)$ may be chosen by the agent at each step. In the traditional Q-learning, each $Q(s, a)$ is put into Q-table. However, with the rapid increase of data dimension, it is challenging to put all $Q(s, a)$ into Q-table.

4.4.6.2 Deep Q-Learning

The rise of deep learning has provided a new tool to overcome the challenges. The most important property of deep learning is that deep networks enable to find the low-dimensional features of high-dimensional data by crafting weights and biases in deep networks. Therefore, many researches have advocated to use deep networks to approximate $Q(s, a)$ instead of Q-table, i.e., $Q(s, a, \omega) \approx Q(s, a)$, where ω is the set of weights and biases in deep networks [124]. This is the core idea of deep Q-learning (DQL).

In order to address the fundamental instability problem of approximating $Q(s, a)$, there are two improvements in DQL, including experience replay and fixed target networks: (1) Experience replay stores the transitions as a set of $\{state, action, reward, state_{next}\}$ in a finite-sized cyclic buffer, and the agent randomly samples batches of them to train deep networks, instead of only the current ones. By this way, the temporal correlations that can adversely affect DQL are broken. (2) Fixed target networks have the same architecture as the evaluated ones, but are kept frozen for a period of time. The evaluated networks are trained in each step to minimize loss function $L(\omega)$ to evaluate real $Q(s, a)$, and $L(\omega)$ is represented as

$$L(\omega) = E[(r + \gamma max_{a'} Q(s', a', \omega^{-}) - Q(s, a, \omega))^2], \tag{4.56}$$

where ω^{-} is the weights and biases set in target networks, and ω is the weights and biases set in evaluated networks. During training, the weights and biases in target networks are updated with evaluated networks periodically. For a comprehensive perspective, we present the workflows of DQL in Fig. 4.34.

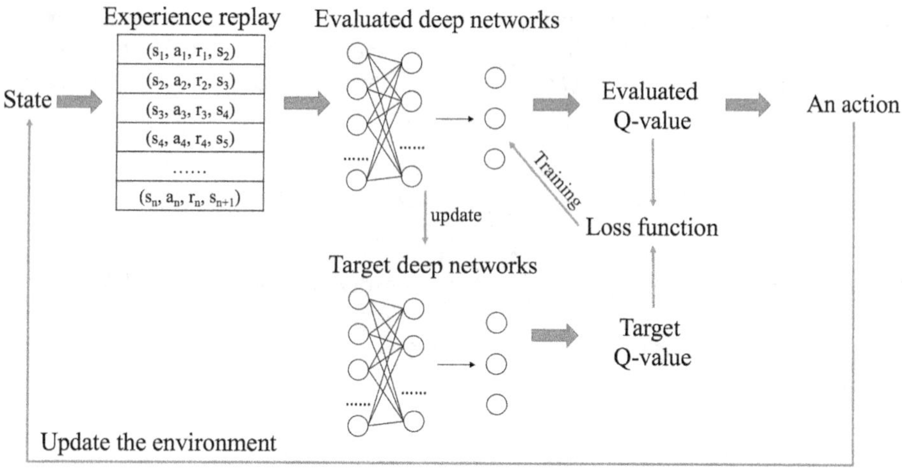

Fig. 4.34 The workflows of DQL

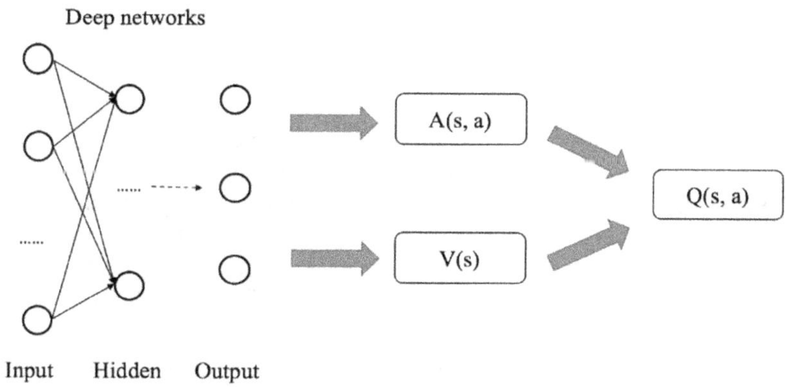

Fig. 4.35 Dueling architecture

4.4.6.3 Dueling Deep Q-Learning Approach

However, for the majority of states in our system, the choice of actions in the agent has no repercussion with what happens, i.e., actions have no relationship with states. According to the work in [58, 125], dueling DQL is more efficient than natural DQL. Based on this approach, some training processes have been carried out in real scenario, such as drive games [125], which are simulated with better performance.

In the dueling DQL, there is another value function, $A(s, a)$, which represents the relative advantage of a action, called state-action value function. Learning $A(s, a)$ is easier to know which action has better consequences. Instead of one single stream following the output layer of deep networks, there are two separate streams in dueling DQL, where one computes state-value function $V(s)$, and another computes state-action value function $A(s, a)$, called dueling architecture as shown in Fig. 4.35.

Finally, these two streams are aggregated as a output $Q(s, a)$. This combination module is denoted as follows:

$$Q(s, a; \omega, \rho, \zeta) = V(s; \omega, \rho) + A(s, a; \omega, \zeta), \tag{4.57}$$

where $V(s; \omega, \rho)$ is a scalar, and $A(s, a; \omega, \zeta)$ is an $|\mathscr{A}|-$dimensional vector. ρ and ζ are the parameters of two separate streams.

Algorithm 4 Dueling DQL

Initialization:
Initialize evaluated deep networks with weights and biases set ω.
Initialize target deep networks with weights and biases set ω^-.
for $k = 1 : K$ do
Reset the environment with a randomly initial observation s_{ini}, and $s(t) = s_{ini}$.
$s(t)! = s_{terminal}$
Select action $a(t)$ based on $\epsilon-$greedy policy.
Obtain immediate reward $r(t)$ and next observation $s(t+1)$.
Store experience $(s(t), a(t), r(t), s(t+1))$ into experience replay memory.
Randomly sample some batches of $(s(i), a(i), r(i), s(i+1))$ from experience replay memory.
Calculate two streams of evaluated deep networks, including $V(s; \omega, \rho)$ and $A(s, a; \omega, \zeta)$, and combine them as $Q(s, a; \omega, \rho, \zeta)$ using (4.59).
Calculate target Q-value $Q_{target}(s)$ in target deep networks:
if s' is $s_{terminal}$
    $Q_{target}(s) = r_s$,
else
    $Q_{target}(s) = r_s + \gamma max_{a'} Q(s', a'; \omega', \rho', \zeta')$.
Train evaluated deep networks to minimize loss function $L(w)$

$$L(\omega, \rho, \zeta) = E[(Q_{target}(s) - Q(s, a; \omega, \rho, \zeta))^2]. \tag{4.58}$$

Every some steps, update target deep networks.
$s(t) \leftarrow s(t+1)$
end while
end for

According to the work in [125], considering the unidentifiability of (4.57), we replace (4.57) as:

$$Q(s, a; \omega, \rho, \zeta) = V(s; \omega, \rho) + \left(A(s, a; \omega, \zeta) - \frac{1}{|\mathscr{A}|} \sum_{a'} A(s, a'; \omega, \zeta)\right). \tag{4.59}$$

As the output of dueling architecture is $Q(s, a)$, it can be trained by many existing algorithms. In this section, we use dueling architecture in natural DQL. Therefore, the pseudocode of dueling DQL is presented in Algorithm 1, where $\epsilon-$greedy policy is used to balance the exploitation and the exploration.

4.4.7 Simulation Results and Discussions

In this subsection, we use computer simulation to evaluate the performance of our proposed scheme. First, we present simulation settings, followed by simulation results and the corresponding discussions.

4.4.7.1 Simulation Settings

In this simulation, hardware environment is a GPU-based server, and this server has 8 GB 1867 MHz LPDDR3, 2 GHz Intel Core i5, and 256 G memory. Software environment is Python 2.7.10 with TensorFlow 1.4.0 [126]. These two kinds of simulation tools have been widely used from business to academic. TensorFlow is a flexible, high-performance serving system, used by machine learning models, and designed for production environments. TensorFlow enables to keep the same server architecture and application program interface (APIs), when different machine learning algorithms are deployed. Thus, it has been widely used to deploy new machine learning algorithms and experiments. We use these real simulation environments, thus we can make sure that the performance of simulation results can estimate and approximate the performance in real scenario.

We assume that there are four BC nodes, two controllers, and two edge computing servers. The trust feature of each node is very safe, safe, medium, compromised, and very compromised, whose transition probability matrix is

$$\mathscr{K} = \begin{bmatrix} 0.5 & 0.15 & 0.125 & 0.12 & 0.105 \\ 0.15 & 0.5 & 0.125 & 0.12 & 0.105 \\ 0.105 & 0.15 & 0.5 & 0.125 & 0.12 \\ 0.105 & 0.12 & 0.125 & 0.5 & 0.15 \\ 0.105 & 0.12 & 0.125 & 0.15 & 0.05 \end{bmatrix}. \tag{4.60}$$

Similarly, the trust feature of each controller can be very safe, safe, medium, compromised, and very compromised. We set the transition probability matrix as

$$\Upsilon = \begin{bmatrix} 0.45 & 0.16 & 0.14 & 0.13 & 0.12 \\ 0.16 & 0.45 & 0.14 & 0.13 & 0.12 \\ 0.12 & 0.16 & 0.45 & 0.14 & 0.13 \\ 0.12 & 0.13 & 0.16 & 0.45 & 0.14 \\ 0.12 & 0.13 & 0.14 & 0.16 & 0.45 \end{bmatrix}. \tag{4.61}$$

We assume the computational capability of each edge computing server is high, medium, low, and very low, and set the transition probability as

$$\Pi = \begin{bmatrix} 0.5 & 0.3 & 0.15 & 0.05 \\ 0.3 & 0.5 & 0.15 & 0.05 \\ 0.15 & 0.3 & 0.5 & 0.05 \\ 0.15 & 0.3 & 0.5 & 0.05 \end{bmatrix}. \tag{4.62}$$

The values of the rest of parameters are summarized in Table 4.6. We use TensorBoard to visualize the TensorFlow graph, as shown in Fig. 4.36.

For the performance comparison, there are four schemes simulated:

- Proposed dueling DQL-based scheme with view changes, access selection, and edge computing servers. We call it duelingDQL-based scheme in the remainder of this section. In other words, in duelingDQL-based scheme, the learning agent enables to select more trusted BC node as the primary, more trusted controller to access to the BC system, and edge computing server with more computing capabilities. Thus, this scheme should have the best performance.

Table 4.6 Parameters setting in the simulation

Parameter	Value	Description
θ	8 Mcycles	The required number of CPU cycles to verify one signature
α	0.05 Mcycles	The required number of CPU cycles to verify and generate one MAC
b	1 Mb	The batch size of a block
γ	0.9	The discount factor

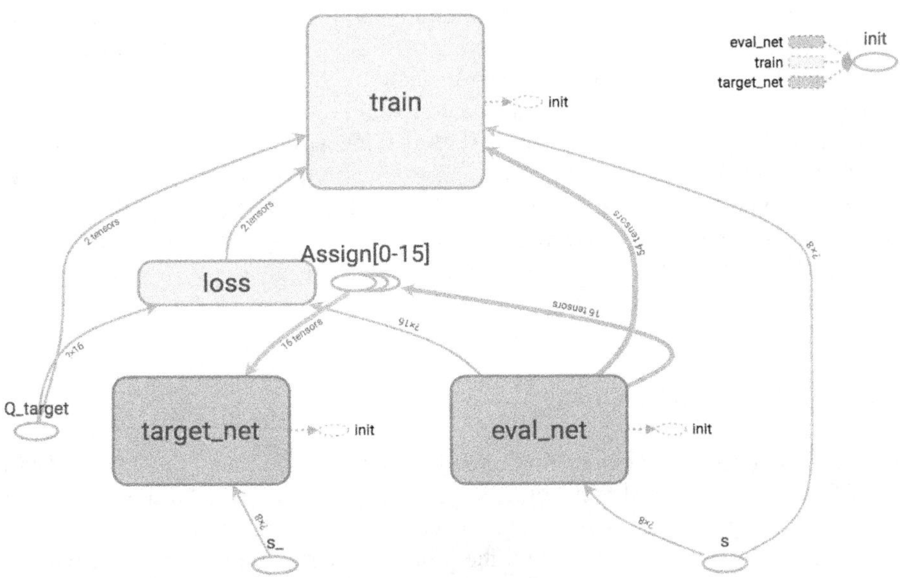

Fig. 4.36 The visualized TensorFlow graph in TensorBoard

- Proposed dueling DQL-based scheme with view changes, edge computing servers, but without access selection. We call it dueling DQL-based scheme without controller choice in the remainder of this section. That is to say, in this scheme, the learning agent only enables to select more trusted BC node as the primary, and edge computing server with more computing capabilities, but controllers access to the BC system randomly. Thus, some malicious controllers may issue incorrect transactions to slow down the performance. Compared with the duelingDQL-based scheme, the advantage of selecting trusted controller can be indicated.
- Proposed dueling DQL-based scheme with access selection, edge computing servers, but with the traditional view change protocol in [107]. We call it duelingDQL-based scheme without node choice in the remainder of this section. In other words, in this scheme, the learning agent only selects more trusted controller to access to the BC system, and edge computing server with more computing capabilities, but using the traditional view change protocol without considering historical trust reputation. Thus, some malicious nodes may be selected as the primary to weaken the throughput. Compared with the duelingDQL-based scheme, the merit of selecting trusted BC node can be shown.
- Proposed dueling DQL-based scheme with view changes, access selection, but without edge computing servers, which only uses the local computing capabilities in the BC system. We call it duelingDQL-based scheme without computation offloading in the remainder of this section. That is to say, the learning agent only selects more trusted BC node as the primary, and more trusted controller to access to the BC system, but only using the local computing capabilities. Without the assistance of edge computing servers, it is difficult to do cryptographic tasks only by local computation, which may slow down the performance. Compared with the duelingDQL-based scheme, the advantage of using edge computing servers can be indicated.
- Existing scheme with traditional view changes, without access selection, and with local computational capabilities. We call it existing scheme in the remainder of this section. The dueling DQL approach is not employed in this scheme. This scheme allows controllers to access to the BC system randomly, uses the traditional view change protocol and local computing capabilities. Compared with the duelingDQL-based scheme, the advantage of using the dueling DQL approach can be shown.

4.4.7.2 Simulation Results

Figure 4.37 shows the relationship between training episodes and the throughput of the BC system under different schemes. Each point is the average throughput per episode. The agent runs in AdamOptimizer [127] with the learning rate of $1e^{-5}$. As we can see from this figure, with the joint consideration of node's trust feature, controller's trust feature, and offloading the computation task to edge computing servers, the BC system has the better performance. The reason is that the more

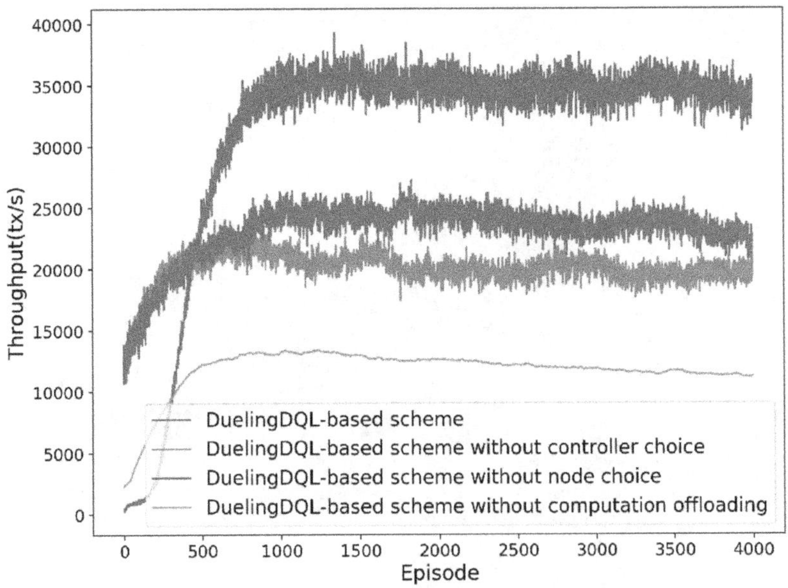

Fig. 4.37 Training curves tracking the throughput of the BC system under different schemes

trusted node is less possible to slow the system performance down, the more trusted controller issues the higher fraction of correct transactions, and with the help of edge computing servers, computation tasks can be executed more quickly. This figure shows the convergence performance of dueling DQL as well. At the beginning of learning and training, dueling DQL takes some trials and errors. With the increase of episodes, the throughput turns to be stable, which means the agent has learned the optimal policies to maximize the long-term rewards.

In addition, Fig. 4.38 shows the relationship between the learning loss in (4.56) and the training steps of DuelingDQL-based scheme, when the agent runs in the same parameters as above. At the beginning of learning, deep networks have no knowledge of the uncertain environment, and with the increase of new experiences, the learning loss is higher and higher. When the cyclic buffer of experiences in dueling DQL is full, the agent has some knowledge of the environment, which leads to the decrease of the learning loss. Such increasing and decreasing of the learning loss indicate the effectiveness of deep networks.

Figure 4.39 shows the relationship between training episodes and the throughput under different learning rates in Dueling DQL-based scheme. As we can see from this figure, the learning rate has effects on the convergence performance. The learning rate means the length of learning step to minimize the loss function. The bigger learning rate denotes the longer learning step. The longer learning steps are likely to miss the global optimum, which leads to the highly scaled curves when learning rates are 0.01 and 0.001. The shorter learning steps may lead to the slower convergence speed, because more steps are necessary to achieve the

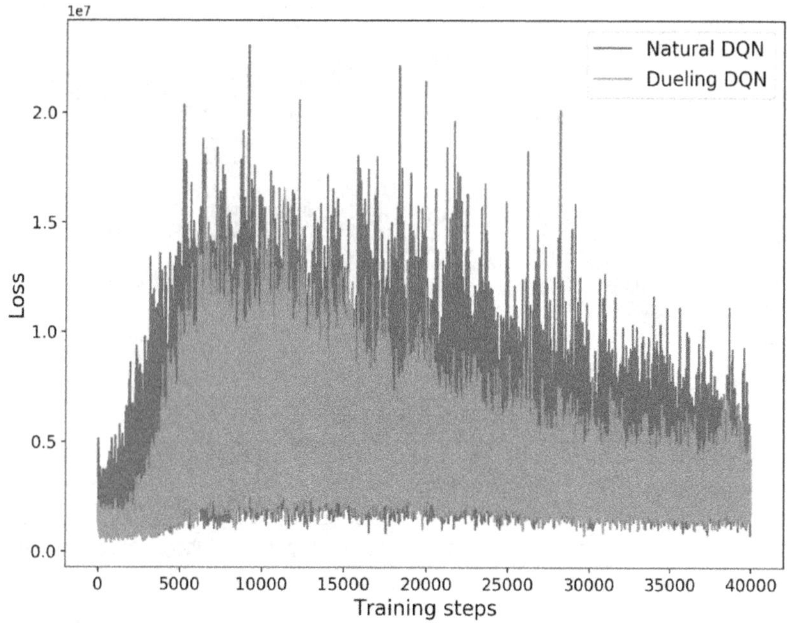

Fig. 4.38 Training curves tracking the learning loss under DuelingDQL-based scheme

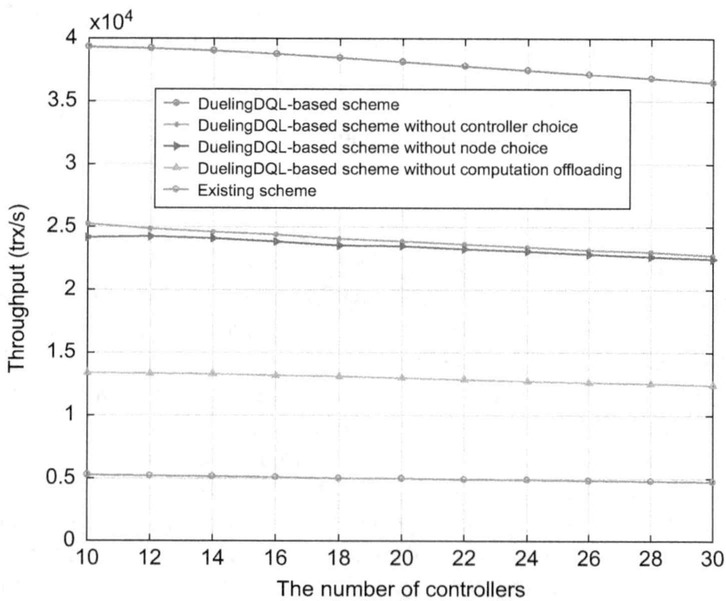

Fig. 4.39 Training curves tracking the throughput of the BC system under different learning rates

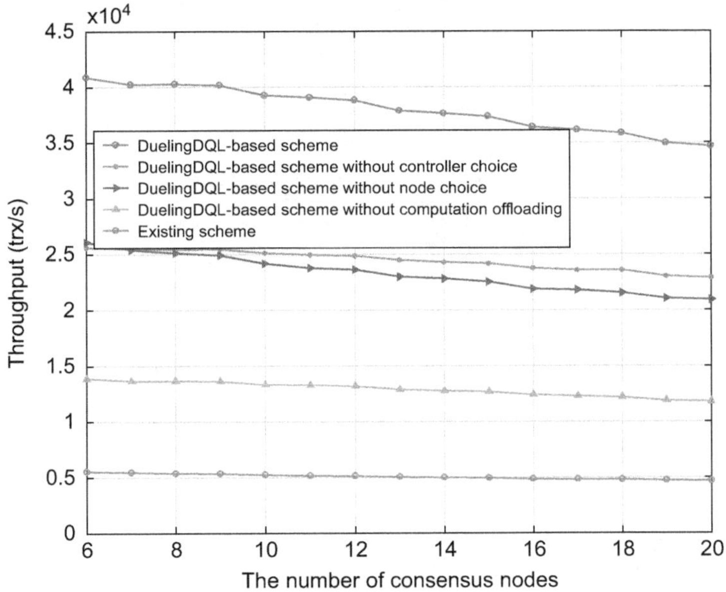

Fig. 4.40 Training curves tracking the learning loss of natural DQL and dueling DQL

global optimum. Compared with two curves of blue and orange, although the orange one has the faster convergence speed, its curve is unstable after the convergence. Therefore, we choose the learning rate as $1e^{-5}$ in the simulation. Because its convergence speed is acceptable and it has better learning stability.

Figure 4.40 shows the learning loss of natural DQL and dueling DQL. As we can see, the learning loss in dueling DQL decreases more quickly than natural DQL, which indicates dueling DQL has better learning effectiveness. The reason is that in our BC system, the choices of which node is the primary, which controller can access to the BC system, and which edge computing server should execute the computation tasks have no relationship with states. Learning which action has better consequence is more efficient than learning which state is better. In dueling DQL, one stream learns state-action value function $A(s, a)$, which is more useful to help the agent make the good choices. Therefore, the learning loss in dueling DQL decreases more fast than natural DQL.

After the effective training of deep networks, we use them in the following simulations. Figure 4.41 shows the relationship between the number of controllers and the system throughput under different schemes. This figure also can be used to show the performance of these schemes in large real SDIIoT environment, where up to 30 controllers are considered in this simulation. As the increase of the number of controllers, the throughput of the BC system decreases. The reason is that more controllers need more computational operations about verifying signatures, MACs. But with the joint consideration of trust features of controllers and nodes, as well as

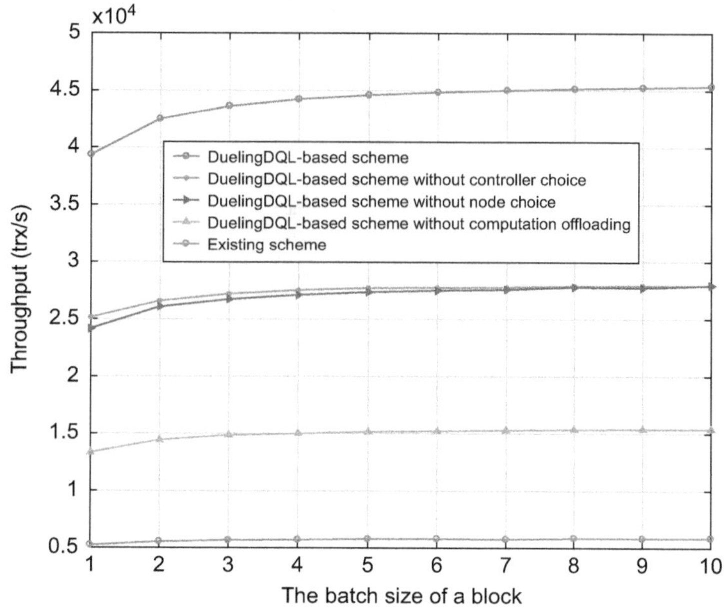

Fig. 4.41 The throughput versus the number of controllers under different schemes

using edge computing servers, our proposed scheme, as shown in the blue curve, has better performance. Thus, we can see our proposed scheme still has better performance in large SDIIoT environment.

Figure 4.42 shows the relationship between the number of BC nodes and the system throughput under different schemes. As we can see, more nodes lead to the less system throughput. The reason is that with the increase of the number of nodes, more signatures and MACs need to be verified and generated, which need more CPU cycles so as to decrease the system throughput. But, the performance of our proposed scheme is still the best.

Figure 4.43 shows the relationship between the batch size of a block and the system throughput. The bigger block enables to contain more transactions so as to synchronize more local network events among controllers, which increases the system throughput. As we can see, our scheme also has the better performance.

4.5 Summary

In this chapter, we discuss the main challenge of network control and introduced several machine learning based algorithms. We first present an energy-aware multi-controller placement scheme as well as a latency-aware resource management model for the SDWN. Then, we present a novel controller mind (CM) framework to

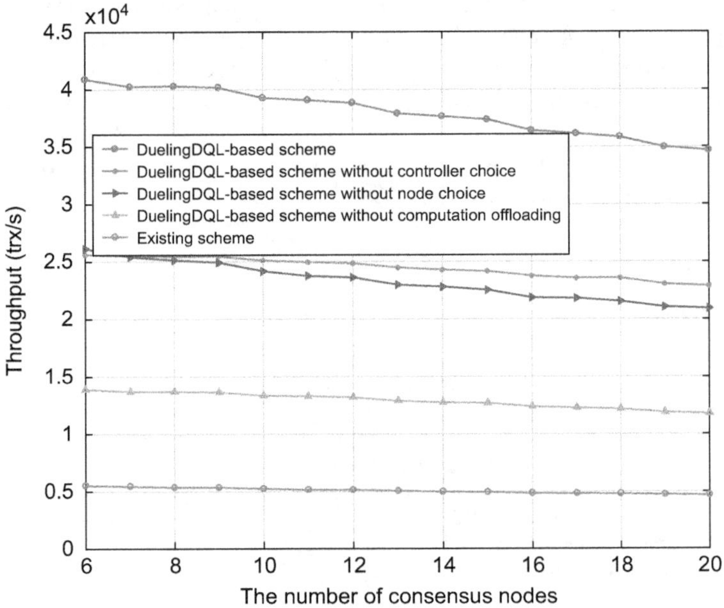

Fig. 4.42 The throughput versus the number of BC nodes under different schemes

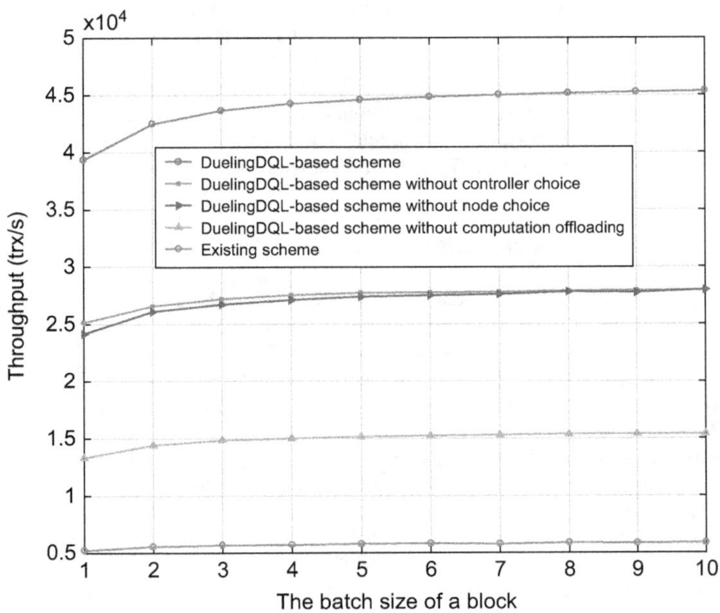

Fig. 4.43 The throughput versus the batch size of a blocks under different schemes

implement automatic management among multiple controllers and propose a novel Quality of Service (QoS) enabled load scheduling algorithm based on reinforcement learning to solve the problem of complexity and pre-strategy in the networks. In addition, we present a Wireless Local Area Networks (WLAN) interference self-optimization method based on a Self-Organizing Feature Map (SOM) neural network model. Finally, we use a novel dueling deep Q-learning approach to solve this joint problem in distributed SDIIoT.

References

1. S. Tamoor-Ul-Hassan, S. Samarakoon, M. Bennis, M. Latva-Aho, and C. S. Hong, "," *IEEE Communications Letters*, vol. 22, no. 1, pp. 137–140, 2018.
2. M. Huang, B. Yu, and S. Li, "Puf-assisted group key distribution scheme for software-defined wireless sensor networks," *IEEE Communications Letters*, vol. 22, no. 2, pp. 404–407, 2018.
3. M. Yue, C. Jiang, H. H. Chen, and R. Yong, "Cooperative device-to-device communications: Social networking perspectives," *IEEE Network*, vol. 31, no. 3, pp. 38–44, 2018.
4. H. Yao, C. Fang, Y. Guo, C. Zhao, H. Yao, C. Fang, Y. Guo, and C. Zhao, "An optimal routing algorithm in service customized 5g networks," *Mobile Information Systems*, vol. 2016, pp. 1–7, 2016.
5. B. Heller, R. Sherwood, and N. McKeown, "The controller placement problem," in *The 1st Workshop on Hot Topics in Software Defined Networks*, Helsinki, Finland, Aug. 2012, pp. 7–12.
6. M. J. Abdel-Rahman, E. A. Mazied, A. MacKenzie, S. Midkiff, M. R. Rizk, and M. El-Nainay, "On stochastic controller placement in software-defined wireless networks," in *IEEE Wireless Communications and Networking Conference (WCNC)*, San Francisco, CA, 2017, pp. 1–6.
7. H. Yao, Y. Hao, A. Zhang, F. Chao, and Y. Guo, "Wlan interference self-optimization using som neural networks," *Concurrency & Computation Practice & Experience*, vol. 29, 2017.
8. R. Grandl, G. Ananthanarayanan, S. Kandula, S. Rao, and A. Akella, "Multi-resource packing for cluster schedulers," *ACM SIGCOMM Computer Communication Review*, vol. 44, no. 4, pp. 455–466, 2015.
9. C. Fang, H. Yao, Z. Wang, W. Wu, X. Jin, and F. R. Yu, "A survey of mobile information-centric networking: Research issues and challenges," *IEEE Communications Surveys & Tutorials*, 2018.
10. V. Jalaparti, P. Bodik, I. Menache, S. Rao, K. Makarychev, and M. Caesar, "Network-aware scheduling for data-parallel jobs: Plan when you can," *ACM SIGCOMM Computer Communication Review*, vol. 45, no. 4, pp. 407–420, 2015.
11. J. Du, C. Jiang, W. Jian, Y. Shui, H. Zhu, and R. Yong, "Resource allocation in space multi-access systems," *IEEE Transactions on Aerospace & Electronic Systems*, vol. 53, no. 2, pp. 598–618, 2017.
12. H. Mao, M. Alizadeh, I. Menache, and S. Kandula, "Resource management with deep reinforcement learning," in *The 15th ACM Workshop on Hot Topics in Networks*, Atlanta, USA, Nov. 2016, pp. 50–56.
13. H. Yao, Q. Chao, F. Chao, C. Xu, and F. R. Yu, "A novel framework of data-driven networking," *IEEE Access*, vol. 4, pp. 9066–9072, 2016.
14. S. Knight, H. X. Nguyen, N. Falkner, R. Bowden, and M. Roughan, "The internet topology zoo," *IEEE Journal on Selected Areas in Communications*, vol. 29, no. 9, pp. 1765–1775, Spe. 2011.
15. C. Jiang, L. Kuang, H. Zhu, R. Yong, and L. Hanzo, "Information credibility modeling in cooperative networks: Equilibrium and mechanism design," *IEEE Journal on Selected Areas in Communications*, vol. 35, no. 2, pp. 432–448, 2017.

16. J. Gao, Y. Xiao, J. Liu, W. Liang, and C. P. Chen, "A survey of communication/networking in smart grids," *Future Generation Comp. Sys.*, vol. 28, no. 2, pp. 391–404, 2012.
17. P. Si, H. Yu, H. Yao, R. Yang, and Y. Zhang, "Dave: Offloading delay-tolerant data traffic to connected vehicle networks," *IEEE Transactions on Vehicular Technology*, vol. 65, no. 6, pp. 3941–3953, 2016.
18. N. McKeown, T. Anderson, H. Balakrishnan, G. Parulkar, L. Peterson, J. Rexford, S. Shenker, and J. Turner, "Openflow: enabling innovation in campus networks," *ACM SIGCOMM Comp. Comm. Review*, vol. 38, no. 2, pp. 69–74, 2008.
19. L. Xu, C. Jiang, Y. Qian, Y. Zhao, J. Li, and Y. Ren, "Dynamic privacy pricing: A multi-armed bandit approach with time-variant rewards," *Trans. Info. For. Sec.*, vol. 12, no. 2, pp. 271–285, Feb. 2017.
20. Y. He, F. R. Yu, N. Zhao, H. Yin, and R. C. Qiu, "Big data analytics in mobile cellular networks," *IEEE Access*, vol. 4, pp. 1985–1996, 2017.
21. S. Al-Rubaye, E. Kadhum, Q. Ni, and A. Anpalagan, "Industrial internet of things driven by SDN platform for smart grid resiliency," *IEEE Internet of Things Journal*, 2017.
22. M. Karakus and A. Durresi, "A survey: Control plane scalability issues and approaches in software-defined networking (SDN)," *Comp. Net.*, vol. 112, pp. 279–293, 2017.
23. C. Jiang, H. Zhang, H. Zhu, R. Yong, V. C. M. Leung, and L. Hanzo, "Information-sharing outage-probability analysis of vehicular networks," *IEEE Transactions on Vehicular Technology*, vol. 65, no. 12, pp. 9479–9492, 2016.
24. H. Yao, Q. Wang, L. Wang, P. Zhang, M. Li, and Y. Liu, "An intrusion detection framework based on hybrid multi-level data mining," *International Journal of Parallel Programming*, pp. 1–19, 2017.
25. C. Qiu, C. Zhao, F. Xu, and T. Yang, "Sleeping mode of multi-controller in green software-defined networking," *EURASIP Journal on Wireless Comm. and Net.*, vol. 2016, no. 1, pp. 282–296, Jan. 2016.
26. T. Zhao, H. Yao, C. Fang, J. Zhang, and P. Zhang, "Microblogging sentiment analysis method with the combination of semantic rules and emoticon weighting," *Journal of Chongqing University of Posts & Telecommunications*, 2016.
27. E. Ng, Z. Cai, and A. Cox, "Maestro: A system for scalable openflow control," *Rice University, Houston, TX, USA, TSEN Maestro-Techn. Rep, TR10-08*, 2010.
28. S. Schmid and J. Suomela, "Exploiting locality in distributed sdn control," in *Proc. Conf. ACM SIGCOMM Workshop on Hot Topics in Software Defined Net., Hong Kong, China*, Aug. 2013, pp. 121–126.
29. J. Wang, C. Jiang, H. Zhu, R. Yong, and L. Hanzo, "Network association strategies for an energy harvesting aided super-wifi network relying on measured solar activity," *IEEE Journal on Selected Areas in Communications*, vol. 34, no. 12, pp. 3785–3797, 2016.
30. C. Liang, F. Yu, H. Yao, and H. Zhu, "Virtual resource allocation in information-centric wireless virtual networks," *IEEE Transactions on Vehicular Technology*, vol. 65, no. 12, pp. 1–1, 2016.
31. H. Yao, B. Zhang, P. Zhang, and M. Li, "A novel kernel for text classification based on semantic and statistical information," *Computing and Informatics*, vol. 37, no. 4, pp. 992–1010, 2018.
32. M. Canini, P. Kuznetsov, D. Levin, and S. Schmid, "A distributed and robust sdn control plane for transactional network updates," in *Proc. Conf. IEEE INFOCOM'15, Hong Kong, China*, Apr. 2015, pp. 190–198.
33. S. Civanlar, M. Parlakisik, A. M. Tekalp, B. Gorkemli, B. Kaytaz, and E. Onem, "A qos-enabled openflow environment for scalable video streaming," in *Pro. Conf. GLOBECOM'10 Workshops, Miami, USA*, Dec. 2010, pp. 351–356.
34. N. T. Hai and D.-S. Kim, "Efficient load balancing for multi-controller in SDN-based mission-critical networks," in *Pro. Conf. Industrial Informatics'16, Poitiers, France*, Jul, 2016, pp. 420–425.
35. P. Zhang, H. Yao, F. Chao, and Y. Liu, "Multi-objective enhanced particle swarm optimization in virtual network embedding," *Eurasip Journal on Wireless Communications & Networking*, vol. 2016, no. 1, p. 167, 2016.

36. M. Yue, C. Jiang, X. Lei, R. Yong, and H. Zhu, "User association in heterogeneous networks: A social interaction approach," *IEEE Transactions on Vehicular Technology*, vol. 65, no. 12, pp. 9982–9993, 2016.
37. P. Zhang, H. Yao, and Y. Liu, "Virtual network embedding based on computing, network, and storage resource constraints," *IEEE Internet of Things Journal*, vol. 5, no. 5, pp. 3298–3304, 2018.
38. F. Xu, H. Yao, C. Zhao, and Q. Chao, "Towards next generation software-defined radio access network–architecture, deployment, and use case," *Eurasip Journal on Wireless Communications & Networking*, vol. 2016, no. 1, p. 264, 2016.
39. L. Meng, F. R. Yu, P. Si, E. Sun, and H. Yao, "Random access and virtual resource allocation in software-defined cellular networks with machine-to-machine (m2m) communications," *IEEE Transactions on Vehicular Technology*, vol. 66, no. 7, pp. 6399–6414, 2017.
40. J. Wang, C. Jiang, T. Q. S. Quek, X. Wang, and R. Yong, "The value strength aided information diffusion in socially-aware mobile networks," *IEEE Access*, vol. 4, pp. 3907–3919, 2016.
41. H. Yao, F. Chao, Q. Chao, C. Zhao, and Y. Liu, "A novel energy efficiency algorithm in green mobile networks with cache," *Eurasip Journal on Wireless Communications & Networking*, vol. 2015, no. 1, pp. 1–9, 2015.
42. J. Wang, C. Jiang, B. Zhi, T. Q. S. Quek, and R. Yong, "Mobile data transactions in device-to-device communication networks: Pricing and auction," *IEEE Wireless Communications Letters*, vol. 5, no. 3, pp. 300–303, 2017.
43. H. Yao, C. Qiu, C. Zhao, and L. Shi, "A multicontroller load balancing approach in software-defined wireless networks," *International Journal of Distributed Sensor Networks*, vol. 2015, no. 2, p. 10, 2015.
44. J. Wang, C. Jiang, H. Zhang, R. Yong, and V. C. M. Leung, "Aggressive congestion control mechanism for space systems," *IEEE Aerospace & Electronic Systems Magazine*, vol. 31, no. 3, pp. 28–33, 2017.
45. H. Yao, C. Qiu, C. Zhao, and L. Shi, "A multicontroller load balancing approach in software-defined wireless networks," *International Journal of Distributed Sensor Networks*, vol. 11, no. 10, pp. 41–49, 2015.
46. X. Lei, C. Jiang, R. Yong, and H. H. Chen, "Microblog dimensionality reduction—a deep learning approach," *IEEE Transactions on Knowledge & Data Engineering*, vol. 28, no. 7, pp. 1779–1789, 2016.
47. However, "Modeling energy-delay tradeoffs in single base station with cache," *International Journal of Distributed Sensor Networks*, vol. 2015, pp. 1–5, 2015.
48. Y. Shen, C. Jiang, T. Q. S. Quek, and R. Yong, "Location-aware green communication design: Exploration and exploitation on energy," *IEEE Wireless Communications*, vol. 23, no. 2, pp. 46–52, 2016.
49. H. Yao, H. Tao, C. Zhao, X. Kang, and Z. Liu, "Optimal power allocation in cognitive radio based machine-to-machine network," *Eurasip Journal on Wireless Communications & Networking*, vol. 2014, no. 1, p. 82, 2014.
50. J. Qi and D. Wu, "Green energy management of the energy internet based on service composition quality," *IEEE Access*, 2018.
51. J. Du, C. Jiang, Q. Yi, H. Zhu, and R. Yong, "Resource allocation with video traffic prediction in cloud-based space systems," *IEEE Transactions on Multimedia*, vol. 18, no. 5, pp. 820–830, 2016.
52. Y. Zhang, R. Yu, M. Nekovee, Y. Liu, S. Xie, and S. Gjessing, "Cognitive machine-to-machine communications: visions and potentials for the smart grid," *IEEE Net.*, vol. 26, no. 3, 2012.
53. S. Maharjan, Q. Zhu, Y. Zhang, S. Gjessing, and T. Basar, "Dependable demand response management in the smart grid: A stackelberg game approach," *IEEE Trans. on Smart Grid*, vol. 4, no. 1, pp. 120–132, 2013.
54. W. Xia, Y. Wen, C. H. Foh, D. Niyato, and H. Xie, "A survey on software-defined networking," *IEEE Comm. Surveys & Tutorials*, vol. 17, no. 1, pp. 27–51, 2015.

55. X. Lei, C. Jiang, Y. Shen, T. Q. S. Quek, H. Zhu, and R. Yong, "Energy efficient d2d communications: a perspective of mechanism design," *IEEE Transactions on Wireless Communications*, vol. 15, no. 11, pp. 7272–7285, 2016.
56. R. Chaudhary, G. S. Aujla, S. Garg, N. Kumar, and J. J. Rodrigues, "SDN-enabled multi-attribute-based secure communication for smart grid in IIoT environment," *IEEE Trans. on Industrial Infor.*, vol. 14, no. 6, pp. 2629–2640, 2018.
57. A. Montazerolghaem, M. H. Yaghmaee, and A. Leon-Garcia, "OpenAMI: Software-defined AMI load balancing," *IEEE Internet of Things Journal*, vol. 5, no. 1, pp. 206–218, 2018.
58. Y. J. Liu, T. Huang, J. Zhang, J. Liu, H. P. Yao, and R. C. Xie, "Service customized networking," *Journal on Communications*, 2014.
59. Y. Zhang, R. Yu, S. Xie, W. Yao, Y. Xiao, and M. Guizani, "Home M2M networks: architectures, standards, and QoS improvement," *IEEE Comm Mag.*, vol. 49, no. 4, 2011.
60. C. Li, S. Tsao, M. C. Chen, Y. Sun, and Y. Huang, "Proportional delay differentiation service based on weighted fair queuing," in *Pro. Conf. Comp. Comm. and Net., Las Vegas, USA*, Oct. 2000, pp. 418–423.
61. SDNCTC, http://www.sdnctc.com/.
62. H. Zhang, C. Jiang, R. Q. Hu, and Q. Yi, "Self-organization in disaster resilient heterogeneous small cell networks," *IEEE Network*, vol. 30, no. 2, pp. 116–121, 2016.
63. http://www.sdnctc.com/download/resource_download/id/6.
64. H. Yao, C. Zhao, and Z. Zhou, "Location based spectrum sensing evaluation in cognitive radio networks," *Eurasip Journal on Wireless Communications & Networking*, vol. 2011, no. 1, pp. 1–7, 2011.
65. Q. Zhang, A. Riska, W. Sun, E. Smirni, and G. Ciardo, "Workload-aware load balancing for clustered web servers," *IEEE Trans. on Parallel and Distributed Sys.*, vol. 16, no. 3, pp. 219–233, 2005.
66. C System, Inc., "Cisco Visual Networking Index: Global Mobile Data Traffic Forecast Update, 2015–2020 [J]", White Paper Cisco Systems Inc. [EB/OL], Feb. 2016.
67. Teuvo Kohonen and Timo Honkela, "Kohonen Network" [J], *Scholarpedia*, pp. 83–100, Jan. 2007. doi:10.4249/scholarpedia.1568
68. RU Qiang and RONG Meng-tian, "Research on IEEE 802.11b WLAN Adjacent Channel Interference", *Information Technology,* vol. 12, pp. 15–17, 2017.
69. Kim K H and Kang G S, "Self-Reconfigurable Wireless Mesh Networks", *IEEE/ACM Transactions on Networking (TON),* vol. 19, no. 2, pp. 393–404, Apr. 2011. doi:10.1109/TNET.2010.2096431
70. Silva M W R D and De Rezende J F, "TDCS: A New Mechanism for Automatic Channel Assignment for Independent IEEE 802.11 Networks", *Ad Hoc Networking Workshop*, 8th IFIP Annual Mediterranean, pp. 27–33, 2009.
71. Leong Yeng Weng, Jamaludin Bin Omar, Yap Keem Siah, Izham Bin Zainal Abidin, and Syed Khaleel Ahmed, "Improvement of ANN-BP by Data Pre-Segregation Using SOM", *IEEE International Conference on Computational Intelligence for Measurement Systems and Application (CIMSA 2009),* pp.175–178, May 11–13, 2009. doi: 10.1109/CIMSA.2009.5069941
72. Chr. von der Malsburg, "Self-organization of Orientation Sensitive Cells in the Striate Cortex", *Journal of High Energy Physics*, vol.4, no. 2, pp, 85–100, Jun. 1973. doi: 10.1007/BF00288907
73. Elsawy, Hesham, E. Hossain and I. K. Dong, "HetNets with Cognitive Small Cells: User Offloading and Distributed Channel Access Techniques", *IEEE Communications Magazine*, vol. 51, no. 6, pp. 28–36, Jun. 2013. doi: 10.1109/MCOM.2013.6525592
74. Mangiameli, P., Chen, S. K., & West, D. (1996), "A Comparison of SOM Neural Network and Hierarchical Clustering Methods", *European Journal of Operational Research*, vol. 93, no. 2, pp. 402–417, Sep. 6th, 1996. doi:10.1016/0377-2217(96)00038-0
75. Owsley, L., Atlas, L., and Bernard, G. (1996), "Self-Organizing Feature Maps with Perfect Organization", *International Conference on Acoustics*, vol. 6, pp. 3557–3560, May. 7th-10th, 1996. doi: 10.1109/ICASSP.1996.550797

76. Chip-Hong C, Pengfei X, Rui X, et al, "New Adaptive Color Quantization Method Based on Self-Organizing Maps" [J], *IEEE Transactions on Neural Networks*, vol. 16, no. 1, pp. 237–249, Feb. 2005. doi: 10.1109/TNN.2004.836543
77. Jian Xianzhong, Cao Shujian, Guo Qiang, "Segmentation of CAPTCHA characters based on self organizing maps and Voronoi", *Application Research of Computers*, Vol.32, No.9, pp. Sep.2015. doi: 2015, **32**(9):2857–2861.
78. 3GPP TS 32.500, "Telecommunication Management; Self-Organizing Networks (SON); Concepts and requirements", Jul. 2008.
79. Gao Y, Wei Y, Fu D, et al, "Research and Application of a New Artificial Immune Algorithm Which Based on SOM Neural Network", *2006 IEEE International Conference on Networking, Sensing and Control,* pp. 1080–1083, 2006. doi: 10.1109/ICNSC.2006.1673302
80. Valero, S., Aparicio, J., Senabre, C., Ortiz, M., Sancho, J., and Gabaldon, A., "Comparative Analysis of Self Organizing Maps vs. Multilayer Perceptron Neural Networks for Short-Term Load Forecasting", *Modern Electric Power Systems (MEPS), 2010 Proceedings of the International Symposium,* pp. 1–5, Sept. 20th – 22nd, 2010.
81. Fang, C., Yu, F. R., Huang, T., & Liu, J., "A Distributed Energy-Efficient Algorithm in Green Content-Centric Networks", *IEEE International Conference on Communications(IOC),* pp. 5546–5551, June 8th-12th, 2015. doi: 10.1109/ICC.2015.7249206
82. Xiaodong X U, Zhang H, Dai X, et al, "SDN Based Next Generation Mobile Network with Service Slicing and Trials" [J], *Wireless Communication over Zigbee for Automotive Inclination Measurement China Communications*, vol. 11, no. 2, pp. 65–77, Feb. 2014. doi: 10.1109/CC.2014.6821738
83. Hai Z, Ding L and Xiang L, "Networking Scientific Resources in the Knowledge Grid Environment: Research Articles" [J], *Concurrency & Computation Practice & Experience,* vol. 19, no, 7, pp. 1087–1113, May. 2007. doi: 10.1002/cpe.1094
84. Chao Fang, F. Richard Yu, Tao Huang, Jiang Liu, and YunJie Liu, "Energy-Efficient Distributed In-Network Caching for Content-Centric Networks", *Global Internet Symposium,* pp. 91–96, 2014.
85. ZHOU Kaili and KANG Yaohong, "Neural Network Model and its MATLAB Simulation Program Design (Chinese Edition)", *Tsinghua University Press,* 2005. ISBN: 9787302108290
86. Z Wang, X Wang, L Liu and M Huang, "Optimal State Feedback Control for Wireless Networked Control Systems with Decentralized Controllers", *Let Control Theory and Applications,* vol. 9, no. 6, pp. 852–862, Apr. 2015. doi: 10.1049/iet-cta.2014.0418
87. Yang C T, Liu J C, Ranjan R, et al, "On Construction of Heuristic QoS Bandwidth Management in Clouds" [J], *Concurrency & Computation Practice & Experience*, vol. 15, no. 18, pp. 2540–2560, Dec. 2013. doi: 10.1002/cpe.3090
88. H Casanova, J Dongarra and DM Doolin, "Java Access to Numerical Libraries", *Concurrency Practice & Experience*, vol. 9, no. 11, pp. 1279–1291, Nov. 1997. doi:10.1002/(SICI)1096-9128(199711)9:11<1279::AID-CPE339>3.0.CO;2-E
89. Chao Fang, F. Richard Yu, Senior Member, IEEE, Tao Huang, Jiang Liu, and Yunjie Liu, "A Survey of Green Information-Centric Networking: Research Issues and Challenges", *IEEE Communication Surveys & Tutorials,* vol. 17, no. 3, pp. 1455–1472, July 2015. doi: 10.1109/COMST.2015.2394307
90. Chao Fang, F. Richard Yu, Tao Huang, Jiang Liu, and Yunjie Liu, "A Survey of Energy-Efficient Caching in Information-Centric Networking", *IEEE Communications Magazine,* vol. 52, no. 11, pp. 122–129, Nov. 2014. doi: 10.1109/MCOM.2014.6957152
91. J.-Q. Li, F. R. Yu, G. Deng, C. Luo, Z. Ming, and Q. Yan, "Industrial internet: A survey on the enabling technologies, applications, and challenges," *IEEE Comm. Surveys & Tutorials*, vol. 19, no. 3, pp. 1504–1526, 2017.
92. H. Zhang, C. Jiang, X. Mao, and H. H. Chen, "Interference-limited resource optimization in cognitive femtocells with fairness and imperfect spectrum sensing," *IEEE Transactions on Vehicular Technology*, vol. 65, no. 3, pp. 1761–1771, 2016.
93. L. Cui, F. R. Yu, and Q. Yan, "When big data meets software-defined networking: SDN for big data and big data for SDN," *IEEE Net.*, vol. 30, no. 1, pp. 58–65, 2016.

94. J. Wan, S. Tang, Z. Shu, D. Li, S. Wang, M. Imran, and A. V. Vasilakos, "Software-defined industrial internet of things in the context of industry 4.0," *IEEE Sensors Journal*, vol. 16, no. 20, pp. 7373–7380, 2016.
95. H. R. Faragardi, H. Fotohi, T. Nolte, and R. Rahmani, "A cost efficient design of a multi-sink multi-controller WSN in a smart factory," in *Proc. Conf. High Performance Comp. and Comm.*, 2017.
96. X. Lei, C. Jiang, C. Yan, W. Jian, and R. Yong, "A framework for categorizing and applying privacy-preservation techniques in big data mining," *Computer*, vol. 49, no. 2, pp. 54–62, 2016.
97. K. Kaur, S. Garg, G. S. Aujla, N. Kumar, J. J. Rodrigues, and M. Guizani, "Edge computing in the industrial internet of things environment: Software-defined-networks-based edge-cloud interplay," *IEEE Comm. Mag.*, vol. 56, no. 2, pp. 44–51, 2018.
98. N. G. Nayak, F. Dürr, and K. Rothermel, "Incremental flow scheduling and routing in time-sensitive software-defined networks," *IEEE Trans. on Industrial Informatics*, vol. 14, no. 5, pp. 2066–2075, 2018.
99. J. Wang, C. Jiang, Z. Han, Y. Ren, and L. Hanzo, "Network association strategies for an energy harvesting aided super-wifi network relying on measured solar activity." *IEEE Journal on Selected Areas in Comm.*, vol. 34, no. 12, pp. 3785–3797, 2016.
100. S. Nakamoto, "Bitcoin: A peer-to-peer electronic cash system," http://bitcoin.org/bitcoin.pdf/, Last Accessed Aug. 2018.
101. F. R. Yu, J. Liu, Y. He, P. Si, and Y. Zhang, "Virtualization for distributed ledger technology (vDLT)," *IEEE Access*, vol. 6, pp. 25 019–25 028, 2018.
102. Y. Shen, C. Jiang, T. Quek, and R. Yong, "Device-to-device assisted communication in cellular network—an energy efficient approach in downlink video sharing scenario," *IEEE Transactions on Wireless Communications*, vol. 15, no. 2, pp. 1575–1587, 2016.
103. N. Group, "The difference between permissionless and permissioned networks," https://medium.com/netis-group-blog/, Last Accessed Aug. 2018.
104. G. Wood, "Ethereum: A secure decentralised generalised transaction ledger," *Ethereum Project Yellow Paper*, vol. 151, pp. 1–32, 2014.
105. C. Cachin and M. Vukolić, "Blockchains consensus protocols in the wild," *arXiv preprint arXiv:1707.01873*, 2017.
106. Y. H. Yang, Y. Chen, C. Jiang, and K. J. R. Liu, "Wireless network association game with data-driven statistical modeling," *IEEE Transactions on Wireless Communications*, vol. 15, no. 1, pp. 512–524, 2016.
107. M. Castro and B. Liskov, "Practical byzantine fault tolerance and proactive recovery," *ACM Trans. on Comp. Sys.*, vol. 20, no. 4, pp. 398–461, 2002.
108. H. Yao, P. Si, R. Yang, and Y. Zhang, "Dynamic spectrum management with movement prediction in vehicular ad hoc networks." *Adhoc & Sensor Wireless Networks*, vol. 32, no. 11, 2016.
109. L. Lamport *et al.*, "Paxos made simple," *ACM Sigact News*, vol. 32, no. 4, pp. 18–25, 2001.
110. C. Cachin, "Architecture of the Hyperledger blockchain fabric," in *Workshop on Distributed Cryptocurrencies and Consensus Ledgers*, 2016, pp. 121–125.
111. J. Guo, X. Liu, C. Jiang, J. Cao, and R. Yong, "Distributed fault-tolerant topology control in cooperative wireless ad hoc networks," *IEEE Transactions on Parallel & Distributed Systems*, vol. 26, no. 10, pp. 2699–2710, 2015.
112. C. Jiang, Y. Chen, Y.-H. Yang, C.-Y. Wang, and K. R. Liu, "Dynamic chinese restaurant game: Theory and application to cognitive radio networks," *IEEE Trans. on Wireless Comm.*, vol. 13, no. 4, pp. 1960–1973, 2014.
113. J. Wang, C. Jiang, H. Zhang, X. Zhang, V. C. Leung, and L. Hanzo, "Learning-aided network association for hybrid indoor LiFi-WiFi systems," *IEEE Trans. on Veh. Tech.*, vol. 67, no. 4, pp. 3561–3574, 2018.
114. C. Jiang, Y. Chen, Y. Gao, and K. R. Liu, "Indian buffet game with negative network externality and non-bayesian social learning," *IEEE Trans. on Sys., Man, and Cybernetics: Sys.*, vol. 45, no. 4, pp. 609–623, 2015.

115. C. Qiu, F. R. Yu, F. Xu, H. Yao, and C. Zhao, "Permissioned blockchain-based distributed software-defined industrial internet of things," in *Globecom Workshops (GC Wkshps), 2018 IEEE*, 2018, pp. 1–7.
116. J. Du, C. Jiang, G. Qiang, M. Guizani, and R. Yong, "Cooperative earth observation through complex space information networks," *IEEE Wireless Communications*, vol. 23, no. 2, pp. 136–144, 2016.
117. C. Qiu, S. Cui, H. Yao, F. Xu, F. R. Yu, and C. Zhao, "A novel QoS-enabled load scheduling algorithm based on reinforcement learning in software-defined energy internet," *Future Generation Comp. Sys.*, 2018.
118. C. Qiu, C. Zhao, F. Xu, and T. Yang, "Sleeping mode of multi-controller in green software-defined networking," *EURASIP Journal on Wireless Commu. and Net.*, vol. 2016, no. 1, p. 282, 2016.
119. H. Yao, C. Qiu, C. Zhao, and L. Shi, "A multicontroller load balancing approach in software-defined wireless networks," *International Journal of Distributed Sensor Net.*, vol. 11, no. 10, p. 454159, 2015.
120. P.-L. Aublin, S. B. Mokhtar, and V. Quéma, "RBFT: Redundant byzantine fault tolerance," in *Proc. Conf. Distributed Comp. Sys.' 13*, 2013, pp. 297–306.
121. A. Clement, E. L. Wong, L. Alvisi, M. Dahlin, and M. Marchetti, "Making byzantine fault tolerant systems tolerate byzantine faults." in *NSDI*, vol. 9, 2009, pp. 153–168.
122. "Hyperledger indy," https://cn.hyperledger.org/projects/hyperledger-indy, Last Accessed Aug. 2018.
123. C. J. Watkins and P. Dayan, "Q-learning," *Machine learning*, vol. 8, no. 3–4, pp. 279–292, 1992.
124. V. Mnih, K. Kavukcuoglu, D. Silver, A. A. Rusu, J. Veness, M. G. Bellemare, A. Graves, M. Riedmiller, A. K. Fidjeland, G. Ostrovski *et al.*, "Human-level control through deep reinforcement learning," *Nature*, vol. 518, no. 7540, p. 529, 2015.
125. Z. Wang, T. Schaul, M. Hessel, H. Van Hasselt, M. Lanctot, and N. De Freitas, "Dueling network architectures for deep reinforcement learning," *arXiv preprint arXiv:1511.06581*, 2015.
126. M. Abadi, A. Agarwal, P. Barham, E. Brevdo, Z. Chen, C. Citro, G. S. Corrado, A. Davis, J. Dean, M. Devin *et al.*, "Tensorflow: Large-scale machine learning on heterogeneous distributed systems," *arXiv preprint arXiv:1603.04467*, 2016.
127. T. M. Chilimbi, Y. Suzue, J. Apacible, and K. Kalyanaraman, "Project Adam: Building an efficient and scalable deep learning training system." in *OSDI*, vol. 14, 2014, pp. 571–582.

Chapter 5
Intelligent Network Resource Management

Resource management problems are ubiquitous in the networking field, such as job scheduling, bitrate adaptation in video streaming and virtual machine placement in cloud computing. In this chapter, we propose a reinforcement learning based dynamic attribute matrix representation (RDAM) algorithm for virtual network embedding. The RDAM algorithm decomposes the process of node mapping into the following three steps: (1) static representation of substrate physical network. (2) dynamic update of substrate physical network. (3) Reinforcement-Learning-Based algorithm. Then, we design and implement a policy network based on reinforcement learning to make node mapping decisions. We use policy gradient to achieve optimization automatically by training the policy network with the historical data based on virtual network requests.

5.1 Virtual Network Embedding Based on RDAM

Network virtualization [1, 3] is regarded as an essential technology for the new generation of Internet architectures. Using network virtualization technology, ISP can create multiple virtual networks on the same substrate physical network, providing users with a variety of customizable end-to-end services [10]. In order to make use of substrate physical network resources effectively, it is necessary to utilize a virtual network embedding algorithm. Virtual network embedding maps virtual requests to specific nodes and links in the substrate physical network. The goal of the virtual network embedding is to make full use of the substrate physical network resources with certain resource constraints.

The virtual network requests contain some constraints, such as the constraints of the attributes of node links, the constraints of request admission, the constraints of dynamically changing request, and so on [6]. Considering multiple constraints simultaneously will make it tough for the virtual network embedding problem

H. Yao et al., *Developing Networks using Artificial Intelligence*, Wireless Networks, https://doi.org/10.1007/978-3-030-15028-0_5

to obtain the optimal node. The optimization process is NP-hard [2, 4, 18, 36]. Therefore, the majority of virtual network embedding algorithms heuristically add assumptions and constraints to reduce the solution space. In this way, the optimal solution can be obtained within an acceptable complexity.

However, the results based on a series of rules and assumptions are not particularly convincing [26, 34]. In addition, the heuristics based on hand-crafted rules are not universal for multiple evaluation metrics. The RDAM algorithm introduced in this section is different from traditional method in three perspectives.

1. Static representation of substrate physical network. The nodes and links information of substrate physical network can be represented by the attribute matrix and the adjacency matrix, respectively. However, these two kind of matrices may be incomplete and noisy. We obtain a robust consensus matrix through spectrum method, and this consensus matrix can effectively represent the substrate physical network.
2. Dynamic update of the substrate network. Substrate physical network needs to be updated after virtual network request arrives. In the virtual network embedding problem, available physical network is changing dynamically at a high frequency [5, 7, 40]. If the consensus matrix is updated every time with the spectrum method, the computational complexity cannot be satisfied in real scenario. We utilize the perturbation theory to capture the changes of nodes and links in the substrate network under continuous time and to complete an efficient method of updating substrate physical network.
3. Reinforcement learning based algorithm. The reinforcement learning agent can effectively discover the relationship between the substrate network representation and virtual network requests, thereby completing an efficient virtual network embedding [11].

5.1.1 System Model and Problem Formulation

5.1.1.1 Description of Virtual Network Embedding

Figuer 5.1 shows the mapping process of two different virtual network requests, where Fig. 5.1a and b are two different virtual requests, and Fig. 5.1c is the substrate network. An undirected graph $G^S = (N^S, L^S, A_N^S, A_L^S)$ is applied to denote the substrate network, where N^S and L^S denote the set of substrate nodes and links, respectively. A_N^S and A_L^S denote the attributes of substrate nodes and links, respectively. Similarly, virtual request can be represented by an undirected graph $G^V = (N^V, L^V, C_N^V, C_L^V)$, where N^V and L^V denote the set of substrate nodes and links, respectively. C_N^V and C_L^V denote the constraints of substrate nodes and links, respectively.

We take Fig. 5.1a as an example. It can be seen from squares boxes that virtual node r1 requires 5 units of computing resources and virtual node r2 requires 10 units

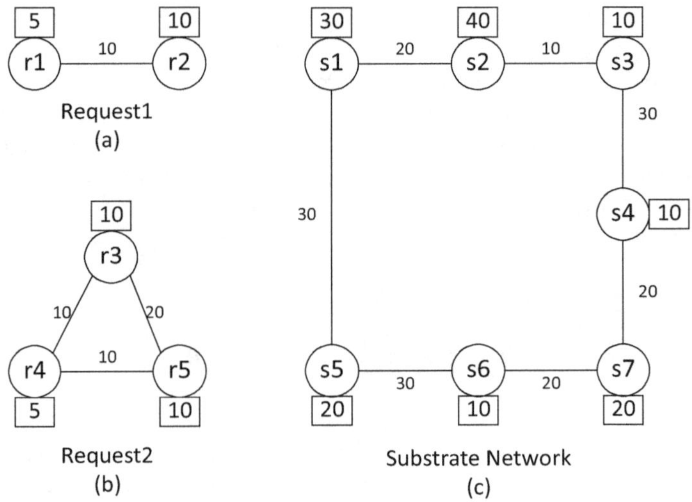

Fig. 5.1 Virtual network request and substrate network

of computing resources. The virtual link between virtual node r1 and virtual node r2 requires 10 units of link resources. Figure 5.1c shows a substrate network that contains 7 substrate nodes from node s1 to node s7. The substrate node s1 holds 30 units of computing resources and the substrate node s2 holds 40 units of computing resources. The link resources between node s1 and node s2 reach 20 units.

The virtual network embedding process can be formulated as a mapping $M : G^V(N^V, L^V) \rightarrow G^S(N^{'}, L^{'})$, where $N^{'} \subset N^S$ and $L^{'} \subset L^S$. As shown in Fig. 5.1, virtual node r1 and virtual node r2 may be mapped to substrate node s1 and substrate node s2, respectively. Then the link request between r1 and r2 is mapped to substrate link between s1 and s2. Note that it is also possible that r1 and r2 are mapped to substrate node s1 and substrate node s3, respectively. Then the link request between r1 and r2 is mapped to two substrate links between substrate node s1 and substrate node s3. Obviously, the latter is not optimal because the link resource between substrate node s2 and substrate node s3 is consumed redundantly.

We use t denote the arrival time of the virtual request in Fig. 5.1a, t_d denote the duration of the virtual request. In the period of t_d, the substrate resources that have been allocated to a request cannot be allocated to other virtual request. Therefore, the virtual network embedding algorithm will have a significant impact on the utilization of substrate resources.

5.1.1.2 Metrics of Virtual Network Embedding

The main goal of virtual network embedding is to map virtual network requests to substrate networks as many as possible [9]. It is beneficial to increase the utilization efficiency of substrate network resources and increase the revenue of infrastructure operators. Most works use three metrics, i.e., long-term average revenue, long-term revenue to cost ratio and long-term acceptance ratio, to measure the utilization efficiency of substrate network resources.

For brevity, in the rest of this part, computing resource is represented by CPU, and link resource is represented by bandwidth. According to the description in [50], the revenue of a virtual network request is

$$R(G^V, t, t_d) = t_d \left[w_c \sum_{n^V \in N^V} CPU(n^V) + w_b \sum_{l^V \in L^V} BW(L^V) \right] \tag{5.1}$$

where w_c and w_b are the weights for CPU and bandwidth, respectively. When a virtual request is allocated, the revenue is generated. No revenue is generated if the request is rejected. The virtual request lasting longer have more revenue.

We define the cost of accepting a virtual request as follows

$$C(G^V, t, t_d) = t_d \left[\sum_{n^V \in N^V} CPU(n^V) + \sum_{l^V \in L^V} \sum_{l^S \in L^S} BW(f_{l^S}^{l^V}) \right] \tag{5.2}$$

where $f_{l^S}^{l^V}$ denotes the total bandwidth l^S allocated to the link request l^V by the virtual network embedding algorithm. In the process of link mapping, a link request l^V may be allocated to multiple links, so the total bandwidth consumption needs to be calculated.

The first metric is long-term average revenue, which is the ratio of revenue to time in infinite time horizon. It measures the overall effect of a virtual network embedding algorithm. The long-term average revenue is defined as

$$Rev = \lim_{T \to \infty} \frac{\sum_{t=0}^{T} R(G^V, t, t_d)}{T}. \tag{5.3}$$

In Eq. (5.3), Rev is the long-term average revenue, $\sum_{t=0}^{T} R(G^V, t, t_d)$ is the total revenue in infinite time horizon. Then it is necessary to achieve a high long-term average revenue with the substrate network resources are consumed less, so we define the long-term revenue to cost ratio

$$RevToCos = \lim_{T \to \infty} \frac{\sum_{t=0}^{T} R(G^V, t, t_d)}{\sum_{t=0}^{T} C(G^V, t, t_d)}. \tag{5.4}$$

In Eq. (5.4), $RevToCos$ is the long-term revenue to cost ratio, $\sum_{t=0}^{T} C(G^V, t, t_d)$ is the total cost in infinite time horizon. Finally, We define the third evaluation metric long-term acceptance ratio $Accept$, which is the ratio of the number of accepted requests to the total number of virtual requests. Defining the long-term acceptance rate contributes to increasing the probability that virtual network requests are accepted, and reduce the number of rejected virtual network requests.

5.1.1.3 Markov Decision Process

We model the virtual network embedding problem as a sequence of decision making instances where an agent receives virtual network requests [12–15]. The agent solves the virtual node mapping and receives a reward. The agent's objective is to maximize this reward.

We use Markov Decision Process (MDP) to model the sequential decision-making problem like [50]. A discrete time MDP M is a quintuple (S, A, R, P, γ), where S is the state space, A is the action space, $R : S \times A \rightarrow R$ is the reward function that assigns real-valued rewards to state-action pairs, and $P : S \times A \times S \rightarrow [0, 1]$ is a transition probability distribution. Therefore, $R(s_t = s, a_t = a)$ is the reward for performing action $a \in A$ in state s_t and

$$P(s', a, s) = Pr(s_{t+1} = s' | a_t = a, s_t = s) \tag{5.5}$$

is the probability of a transition to state s' when selecting action a in state s. The behavior of a decision-making agent is defined by its policy π of selecting actions. The overall return of a given policy π is calculated by

$$R_t = r_{t+1} + \gamma r_{t+2} + \gamma^2 r_{t+3} \ldots = \sum_{k=0}^{\infty} \gamma^k r_{t+k+1}, \tag{5.6}$$

where $0 \leq \gamma^t \leq 1$ is a discount factor, r_{t+1} represents the revenue in time step t+1. The goal of the decision-making agent is to find the policy π^* that maximizes the expected cumulative rewards given an initial state $s1$.

5.1.1.4 Feature Attribute Extraction

In the RDAM algorithm, the description of the substrate node attribute is very important for the reinforcement learning agent to comprehend substrate network and learn mapping rules. The work presented in [20] extracts the following substrate node attributes:

1. Computing Resources (CPU): The computing resource of a substrate node determines its availability. When the computing resource of a substrate node is sufficient, it is more likely to accept node requests.

2. Degree (DEG): The degree denotes the connectivity of substrate network. Nodes with better connectivity are more likely to be occupied by node requests.
3. Sum of bandwidth (SUM^{BW}): The sum of substrate network bandwidth describes connectivity from the perspective of available bandwidth. When a substrate node has access to more bandwidth, mapping a virtual node to it may lead to better link mapping options.
4. Average distance to other host nodes (AVG^{DST}): The fourth attribute of a substrate node is the average distance from it to other already mapped node in the same request. If we select a substrate node close to those already mapped, the cost of bandwidth can be effectively reduced. The combination of mapped nodes with few hops can effectively save the link resources of the substrate network.

After above four attributes are extracted, they need to be normalized to facilitate subsequent node sorting and optimization. The normalization method for CPU, DEG, and SUM^{BW} is to divide the corresponding attribute value by the maximum attribute value in the initial state of substrate physical network. For AVG^{DST}, the initialization mode is

$$\frac{1}{AVG^{(DST)}+1}. \tag{5.7}$$

By concatenate these four attributes together, we get the node's attribute matrix A, i.e.,

$$A = [CPU, DEG, SUM^{(BW}, AVG^{(DST)}]. \tag{5.8}$$

We summarize some notations in the RDAM algorithm. The uppercase symbol represents a matrix (eg, A), the arrow symbol represents a vector (eg, **a**), and the lower case symbol represents a scalar (eg, a). A' denotes the transpose of the matrix A. As we describe the physical network $G^S = (N^S, L^S, A_N^S, A_L^S)$ in Sect. 5.1.1.1, A_N^S represents the attributes of the nodes and A_L^S represents the attributes of the links. Therefore, $A^{(t)}$ denotes an attribute matrix at time step t, and $X^{(t)}$ denotes an adjacency matrix at time step t. In Table 5.1, the definitions of all the symbols used in the RDAM algorithm are given.

5.1.2 RDAM Algorithm

This subsection will mainly introduce the RDAM algorithm, including static representation of substrate network, dynamic update of substrate network, model construction, optimization method and, the overall steps of training and testing.

The RDAM algorithm regards virtual network embedding as two stages: node mapping and link mapping. In the link mapping stage, the nodes already mapped are assigned links by breadth-first traversal. The process of node mapping is described

Table 5.1 Symbols

Notations	Definitions
$A^{(t)}$	Attribute matrix at time step t
$X^{(t)}$	Adjacency matrix at time step t
$A^{(t+1)}$	Attribute matrix at time step t+1
$X^{(t)}$	Adjacency matrix at time step t+1
ΔA	Change of attribute matrix between time steps t and t+1
ΔX	Change of adjacency matrix between time steps t and t+1
n	Number of nodes in substrate network
d	Number of node attributes
k	Embedding dimension for network attributes or structure
l	Final consensus embedding dimension

in detail in Fig.5.2. It is divided into Problem 1 and Problem 2. Problem 1 refers to the problem of static representation of substrate network. Problem 2 refers to the problem of dynamic update of the substrate network.

We utilize the spectrum method in problem 1 to obtain a robust consensus matrix of substrate network. We adopt the matrix perturbation theory in problem 2 to reduce complexity of the static update method.

5.1.2.1 Static Representation of Substrate Network

The substrate network has node information and link information. The node information of substrate network can be represented by an attribute matrix; the link information of substrate network can be represented by an adjacency matrix. We divide the process of static representation of substrate network into two steps. The first step is to reduce the noise from the attribute matrix $A^{(t)}$ and adjacency matrix $X^{(t)}$. We obtain the embedding representation $Y_A^{(t)}$ of the attribute matrix $X^{(t)}$ and the embedding representation $Y_X^{(t)}$ of $A^{(t)}$. In the second step, the final consensus matrix $Y^{(t)}$ is obtained from $Y_X^{(t)}$ and $Y_A^{(t)}$.

Let $D_X^{(t)} \in \mathbb{R}^{n\times n}$ be the degree matrix of $X^{(t)}$, i.e., $D_X^{(t)}(i,i) = \sum_{j=1}^{n} X^{(t)}(i,j)$. Then $L_X^{(t)} = D_X^{(t)} - X^{(t)}$ is a Laplacian matrix. According to spectral theory [21, 22], mapping an n-dimensional matrix to a k-dimensional embedding matrix ($k \ll n$) can effectively reduce the noise in the matrix representation. A universal choice of $Y_X^{(t)} = [y_1, y_2, \ldots, y_n]' \in \mathbb{R}^{n\times k}$ is to minimize the loss function $\frac{1}{2}\sum_{j=1}^{n} X^{(t)}(i,j)\|y_i - y_j\|_2^2$. In the embedding space, the distance between interconnected nodes is closer.

The first step degenerates into a generalized eigen-problem $L_X^{(t)}\mathbf{a} = \lambda D_X^{(t)}\mathbf{a}$ [23]. Assuming $\mathbf{a_1}, \mathbf{a_2}, \ldots, \mathbf{a_n}$ are the eigenvectors of the corresponding eigenvalues $0 = \lambda_1 \leqslant \lambda_2 \leqslant \ldots \leqslant \lambda_n$, it is easy to verify $\lambda_1 = 0$ and the corresponding eigenvector is

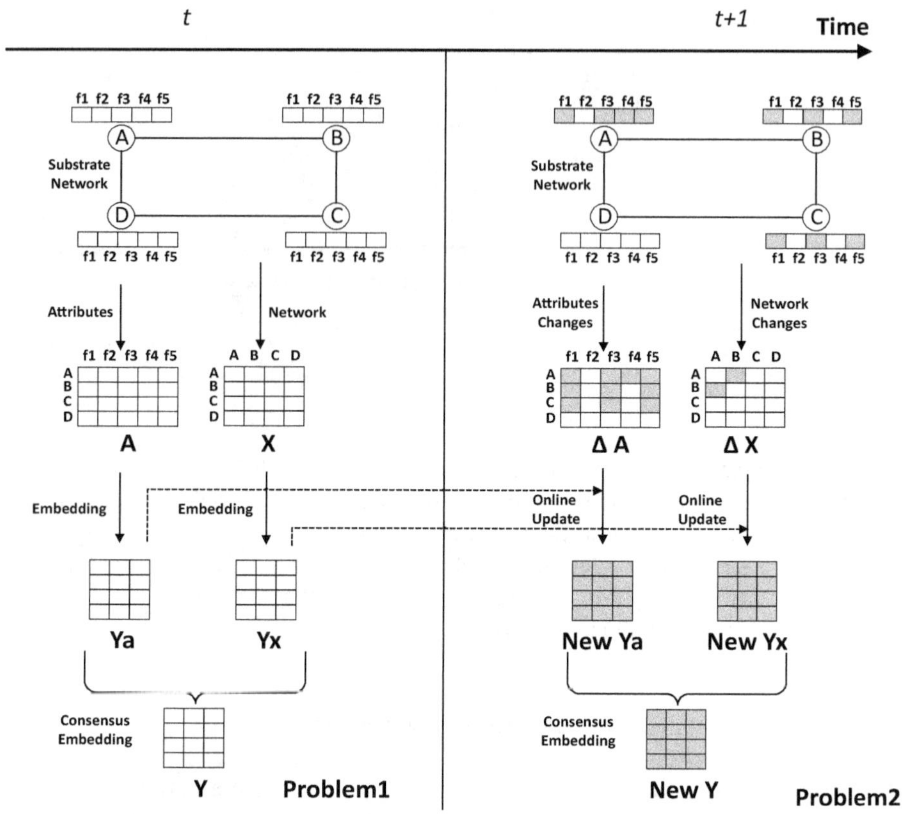

Fig. 5.2 The left side denotes the problem 1 at time step t and the right side denotes the problem 2 at time step t+1. At time step t, we first obtained the attribute matrix $A^{(t)}$ and adjacency matrix $X^{(t)}$ of substrate network. The noise in aforementioned two matrices is eliminated by spectral analysis. We obtain the embedding representation $Y_A^{(t)}$ and $Y_X^{(t)}$ of the attribute matrix and adjacency matrix. Then we get the final consensus matrix through two embedding matrices. Then at time step t+1, the grey grid in substrate network denotes the changed attribute ΔA and changed structure ΔX. We will obtain the new embedding representation $Y_A^{(t)}$ and $Y_X^{(t)}$ from the variation of the attribute matrix and the adjacency matrix. Finally, we get the new consensus matrix $Y^{(t+1)}$

unit vector $\mathbb{1}$. Then the k-dimensional embedding $Y_X^{(t)} \in \mathbb{R}^{n \times k}$ of network structure is given by the top-k eigenvectors starting from $\mathbf{a_2}$, i.e., $Y_X^{(t)} = [\mathbf{a_2}, \mathbf{a_3}, \ldots, \mathbf{a_{k+1}}]$.

Similar to the adjacency matrix, the attribute embedding matrix $Y_A^{(t)}$ can also be obtained in such a way. First we normalize the attribute matrix and obtain the similarity matrix $W^{(t)}$ of the normalized attribute matrix. Then we solve the generalized eigen-problem and obtain the embedding representation of the attribute matrix $Y_A^{(t)}$.

The noise of $X^{(t)}$ and $A^{(t)}$ has been eliminated by computing adjacency embedding matrix $Y_X^{(t)}$ and the attribute embedding matrix $Y_A^{(t)}$. Now we utilize them to seek a consensus matrix. However, these two embedding matrices are obtained in different ways and may not be relevant. In order to capture their interdependency and make them compensate each other, we maximize the correlation of them. We utilize two projection vectors $P_X^{(t)}$ and $P_A^{(t)}$ to maximize the correlation of $Y_X^{(t)}$ and $Y_A^{(t)}$ after projection. The problem is equivalent to

$$\begin{aligned} \max_{P_A^{(t)},P_X^{(t)}} \; & P_A^{(t)'} Y_A^{(t)'} Y_A^{(t)} P_A^{(t)} + P_A^{(t)'} Y_A^{(t)'} Y_X^{(t)} P_X^{(t)} \\ & + P_X^{(t)'} Y_X^{(t)'} Y_A^{(t)} P_A^{(t)} + P_X^{(t)'} Y_X^{(t)'} Y_X^{(t)} P_X^{(t)}, \\ s.t. \quad & P_A^{(t)'} Y_A^{(t)'} Y_A^{(t)} P_A^{(t)} + P_X^{(t)'} Y_X^{(t)'} Y_X^{(t)} P_X^{(t)} = 1. \end{aligned} \tag{5.9}$$

This is a typical optimization problem. Let γ be a Lagrangian operator. Then the value of $P_X^{(t)}$ and $P_A^{(t)}$ can be obtained by setting the derivative of the Lagrangian function w.r.t. $P_X^{(t)}$ and $P_A^{(t)}$ to zero. The expression of the corresponding generalized eigen-problem is

$$\begin{bmatrix} Y_A^{(t)'} Y_A^{(t)} & Y_A^{(t)'} Y_X^{(t)} \\ Y_X^{(t)'} Y_A^{(t)} & Y_X^{(t)'} Y_X^{(t)} \end{bmatrix} \begin{bmatrix} P_A^{(t)} \\ P_X^{(t)} \end{bmatrix} = \gamma \begin{bmatrix} Y_A^{(t)'} Y_A^{(t)} & 0 \\ 0 & Y_X^{(t)'} Y_X^{(t)} \end{bmatrix} \begin{bmatrix} P_A^{(t)} \\ P_X^{(t)} \end{bmatrix}. \tag{5.10}$$

We take the top-l eigenvectors as the consensus matrix of the above generalized eigen-problem. Assuming a projection matrix $P^{(t)} \in \mathbb{R}^{2k \times l}$, then the final consensus matrix is expressed as $Y^{(t)} = [Y_A^{(t)}, Y_X^{(t)}] \times P^{(t)} \in \mathbb{R}^{n \times l}$.

5.1.3 *Dynamic Update of Substrate Network*

In the virtual network embedding process, the substrate network will change when a virtual request is accepted. Therefore, the attribute matrix and adjacency matrix are changing frequently. If we obtained the consensus matrix $Y^{(t)}$ by computing eigenvectors and eigenvalues of generalized eigen-problem at each time step, this method will be time-consuming and it is not practical in large-scale networks. Therefore, we propose to update $Y_A^{(t)}$ and $Y_X^{(t)}$ by matrix perturbation theory, which can reduce the computational complexity.

We propose the dynamical update method based on the fact that the attribute matrix and adjacency matrix do not change too much in two consecutive time steps, as virtual request resources usually do not occupy a large part of substrate resources. We utilize ΔA and ΔX to donate the variation of the attribute matrix and the adjacency matrix within two consecutive time steps. The degree matrix and the Laplacian matrix are given by

$$\begin{aligned} D_A^{(t+1)} &= D_A^{(t+1)} + \Delta D_A, L_A^{(t+1)} = L_A^{(t+1)} + \Delta L_A, \\ D_X^{(t+1)} &= D_X^{(t+1)} + \Delta D_X, L_X^{(t+1)} = L_X^{(t+1)} + \Delta L_X. \end{aligned} \tag{5.11}$$

We take the update method of the embedding matrix of the adjacency matrix as an example to illustrate the specific steps of dynamic update method. According to matrix perturbation theory [24], we have

$$\begin{aligned} &(L_X^{(t)} + \Delta L_X)(\mathbf{a} + \Delta \mathbf{a}) = \\ &\qquad (\lambda + \Delta\lambda)(D_X^{(t)} + \Delta D_X)(\mathbf{a} + \Delta \mathbf{a}). \end{aligned} \tag{5.12}$$

Then, for a certain eigenvalue and eigenvector pair $(\lambda_i, \mathbf{a_i})$, we have

$$\begin{aligned} &(L_X^{(t)} + \Delta L_X)(\mathbf{a}_i + \Delta \mathbf{a}_i) = \\ &\qquad (\lambda_i + \Delta\lambda_i)(D_X^{(t)} + \Delta D_X)(\mathbf{a}_i + \Delta \mathbf{a}_i). \end{aligned} \tag{5.13}$$

The algorithm problem of dynamic update is equivalent to how to obtain the changed pairs $(\Delta\lambda_i, \Delta\mathbf{a_i})$ from the changed Laplacian matrix ΔL and the changed degree matrix ΔD. The way to solve $(\Delta\lambda_i, \Delta\mathbf{a_i})$ is divided into following two parts.

Calculate $\Delta\lambda_i$

By expanding Eq. (5.13), we have

$$\begin{aligned} &L_X^{(t)}\mathbf{a}_i + \Delta L_X\mathbf{a}_i + L_X^{(t)}\Delta\mathbf{a}_i + \Delta L_X\Delta\mathbf{a}_i \\ &= \lambda_i D_X^{(t)}\mathbf{a}_i + \lambda_i\Delta D_X\mathbf{a}_i + \Delta\lambda_i D_X^{(t)}\mathbf{a}_i + \Delta\lambda_i\Delta D_X\mathbf{a_i} \\ &+ (\lambda_i D_X^{(t)} + \lambda_i\Delta D_X + \Delta\lambda_i D_X^{(t)} + \Delta\lambda_i\Delta D_X)\Delta\mathbf{a}_i, \end{aligned} \tag{5.14}$$

where these higher order terms, i.e., $\Delta\lambda_i\Delta D_X\mathbf{a_i}$, $\lambda_i\Delta D_X\Delta\mathbf{a}_i$, $\Delta\lambda_i D_X^{(t)}\Delta\mathbf{a}_i$, and $\Delta\lambda_i\Delta D_X\Delta\mathbf{a}_i$ can be removed as they has narrow effects on the accuracy of a generalized eigen-problem. In addition, since $L_X^{(t)}\mathbf{a} = \lambda D_X^{(t)}\mathbf{a}$, Eq. (5.14) can be simplified as

$$\begin{aligned} &\Delta L_X\mathbf{a}_i + L_X^{(t)}\Delta\mathbf{a}_i \\ &= \lambda_i\Delta D_X\mathbf{a}_i + \Delta\lambda_i D_X^{(t)}\mathbf{a}_i + \lambda_i D_X^{(t)}\Delta\mathbf{a}_i. \end{aligned} \tag{5.15}$$

Next, multiplying both sides of Eq. (5.15) with $\mathbf{a_i}'$, we have

$$\begin{aligned} & a_i' \Delta L_X \mathbf{a}_i + a_i' L_X^{(t)} \Delta \mathbf{a}_i \\ & = a_i' \lambda_i \Delta D_X \mathbf{a}_i + a_i' \Delta \lambda_i D_X^{(t)} \mathbf{a}_i + a_i' \lambda_i D_X^{(t)} \Delta \mathbf{a}_i \end{aligned} \tag{5.16}$$

Since the Laplacian matrix $L_X^{(t)}$ and the degree matrix $D_X^{(t)}$ are symmetric, we have

$$a_i' L_X^{(t)} \Delta \mathbf{a}_i = a_i' \lambda_i D_X^{(t)} \Delta \mathbf{a}_i \tag{5.17}$$

So, Eq. (5.12) becomes

$$a_i' \Delta L_X \mathbf{a}_i = a_i' \lambda_i \Delta D_X \mathbf{a}_i + a_i' \Delta \lambda_i D_X^{(t)} \mathbf{a}_i. \tag{5.18}$$

Finally, the change of eigenvalue $\Delta \lambda_i$ is given by

$$\Delta \lambda_i = \frac{a_i' \Delta L_X \mathbf{a}_i - a_i' \lambda_i \Delta D_X \mathbf{a}_i}{a_i' D_X^{(t)} \mathbf{a}_i}. \tag{5.19}$$

Theorem 5.1 *In the generalized eigen-problem $Av = \lambda B v$, if A and B are both Hermitian matrices and B is positive-semidefinite matrice, the eigenvectors are real, eigenvectors $v_j (i \neq j)$ and B is orthogonal, then, $v_i' B v_j = 0$ and $v_i' B v_i = 1$.*

In virtual network embedding problem: $a_i' D_X^{(t)} a_j = 0 (i \neq j)$ and $a_i' D_X^{(t)} a_i = 1 (i = j)$

Proof Degree matrix $D_X^{(t)}$ and Laplacian matrix $L_X^{(t)}$ are symmetric Hermitian matrices. At the same time, the degree matrix $D_X^{(t)}$ is a semi-positive definite matrix, thus completing the proof. From above proof, we have

$$\Delta \lambda_i = a_i' \Delta L_X \mathbf{a}_i - a_i' \lambda_i \Delta D_X \mathbf{a}_i. \tag{5.20}$$

Calculate $\Delta \mathbf{a_i}$

Since the network structure changes smoothly in continuous time steps, we assume that the perturbation of the eigenvectors consists of the top-k eigenvectors, i.e., $\Delta \mathbf{a_i} = \sum_{j=2}^{k+1} \alpha_{ij} \mathbf{a_j}$, where α_{ij} is the weight of the j-th eigenvector.

By plugging $\Delta \mathbf{a_i} = \sum_{j=2}^{k+1} \alpha_{ij} \mathbf{a_j}$ into Eq. (5.15), we have

$$\begin{aligned} & \Delta L_X \mathbf{a}_i + D_X^{(t)} \sum_{j=2}^{k+1} \alpha_{ij} \lambda_j \mathbf{a_j} \\ & = \lambda_i \Delta D_X \mathbf{a}_i + \Delta \lambda_i D_X^{(t)} \mathbf{a}_i + \lambda_i D_X^{(t)} \sum_{j=2}^{k+1} \alpha_{ij} \mathbf{a_j}. \end{aligned} \tag{5.21}$$

When $p \neq i$, by multiplying both sides with $\mathbf{a_p}'(2 \leqslant p \leqslant k+1, p \neq i)$ on both sides of Eq. (5.21) and using the theorem 1, we have:

$$\begin{aligned}
&\mathbf{a}_\mathbf{p}' \Delta L_X \mathbf{a}_i + \mathbf{a}_\mathbf{p}' D_X^{(t)} \sum_{j=2}^{k+1} \alpha_{ij} \lambda_j \mathbf{a_j} \\
&= \lambda_i \mathbf{a}_\mathbf{p}' \Delta D_X \mathbf{a}_i + \Delta\lambda_i \mathbf{a}_\mathbf{p}' D_X^{(t)} \mathbf{a}_i + \lambda_i \mathbf{a}_\mathbf{p}' D_X^{(t)} \sum_{j=2}^{k+1} \alpha_{ij} \mathbf{a_j} \\
&\rightarrow \mathbf{a}_\mathbf{p}' \Delta L_X \mathbf{a}_i + \alpha_{ip}\lambda_p = \lambda_i \mathbf{a}_\mathbf{p}' \Delta D_X \mathbf{a}_i + \alpha_{ip}\lambda_i.
\end{aligned} \tag{5.22}$$

Therefore, we obtain the weight α_{ip} as

$$\alpha_{ip} = \frac{\mathbf{a}_\mathbf{p}' \Delta L_X \mathbf{a}_i - \lambda_i \mathbf{a}_\mathbf{p}' \Delta D_X \mathbf{a}_i}{\lambda_i - \lambda_p}. \tag{5.23}$$

When $p = i$, from Theorem 5.1, we have

$$(\mathbf{a}_i + \Delta \mathbf{a}_i)'(D_X + \Delta D_X)(\mathbf{a}_i + \Delta \mathbf{a}_i) = 1. \tag{5.24}$$

By ignoring the high order term in Eq. (5.24), we get

$$2\mathbf{a}_i' D_X^{(t)} \Delta \mathbf{a}_i + \mathbf{a}_i' \Delta D_X^{(t)} \mathbf{a}_i = 0, \tag{5.25}$$

namely,

$$\alpha_{ii} = -\frac{1}{2} \mathbf{a}_i' \Delta D_X^{(t)} \mathbf{a}_i. \tag{5.26}$$

Therefore, the change of eigenvalue $\Delta \mathbf{a_i}$ is

$$\begin{aligned}
\Delta \mathbf{a}_i =& -\frac{1}{2} \mathbf{a}_i' \Delta D_X^{(t)} \mathbf{a}_i \mathbf{a}_i \\
&+ \sum_{j=2, j\neq i}^{k+1} \left(\frac{\mathbf{a}_\mathbf{p}' \Delta L_X \mathbf{a}_i - \lambda_i \mathbf{a}_\mathbf{p}' \Delta D_X \mathbf{a}_i}{\lambda_i - \lambda_p}\right) \mathbf{a_j}.
\end{aligned} \tag{5.27}$$

Now, we get the perturbation pairs $(\Delta\lambda_i, \Delta \mathbf{a_i})$. The pseudo code is given in Algorithm 1. At start time(t=1), we obtain the initial embedding attribute matrix and embedding adjacency matrix by spectral method. In addition, we have the initial eigenvalue and eigenvector pairs $(\lambda_i, \mathbf{a_i})$ at start time. The input of the algorithm is $(\lambda_i, \mathbf{a_i})$ at start time.The algorithm computes the perturbation of the Laplacian and degree matrices for each time step. The output is the first k eigenvalues and eigenvector pairs at time step T.

In Algorithm 1, lines 3 and 4 call Eq. (5.20) and (5.27) to calculate the perturbation, and then update the eigenvalues and eigenvectors. The embedding matrix of attribute matrices can be updated in the same way.

Algorithm 1 Updating of embedding matrix for substrate network

---

Input: Top-k eigen-pairs of the generalized eigen-problem $\{(\lambda_2, \overrightarrow{a_2}),(\lambda_3, \overrightarrow{a_3}),\ldots,(\lambda_{k+1}, \overrightarrow{a_{k+1}})\}$ at start time(t=1). Perturbation of the diagonal matrix ΔL_A and Laplacian matrix ΔD_A at each time.

Output: Top-k eigen-pairs $\{(\lambda_2^{(T)}, \overrightarrow{a_2}^{(T)}), (\lambda_3^{(T)}, \overrightarrow{a_3}^{(T)}),\ldots, (\lambda_{k+1}^{(T)}, \overrightarrow{a_{k+1}}^{(T)})\}$ at time T.

1: **for** $t = 1 \to T - 1$ **do**
2:     **for** $i = 2 \to k + 1$ **do**
3:         Calculate the variation of $\Delta\lambda_i$ by Eq. (5.20);
4:         Calculate the variation of $\Delta\overrightarrow{a_i}$ by Eq. (5.27);
5:         $\lambda_i^{(t+1)} = \lambda_i^{(t)} + \Delta\lambda_i$
6:         $\overrightarrow{a_i}^{(t+1)} = \overrightarrow{a_i}^{(t)} + \Delta\overrightarrow{a_i}$
7:     **end for**
8: **end for**

5.1.3.1 Computational Complexity

The computational complexity of static method is $O(n^2k + k^2l)$, where n is number of substrate nodes, k is the dimension of the embedding representation of the adjacency or attribute matrix, and l is the dimension of the final consensus matrix.

Proof For an $n \times n$ square matrix, the computational complexity of solving all the eigenvectors is $O(n^3)$, and the computational complexity of solving the eigenvector corresponding to a certain eigenvalue is $O(n^2)$. In the static method, the computational complexity of getting the embedding representation of the adjacency matrix is $O(n^2k)$, and the computational complexity of the embedding consensus matrix is $O(k^2l)$.

The computational complexity of dynamic update is $O(k^2(n + l + l_a + l_x + d_a + d_x))$, where l_a, l_x, d_a, d_x are the number of non-zero elements in $\Delta L_A, \Delta L_X, \Delta D_A, \Delta D_X$, respectively.

Proof In dynamic update method, the computational complexity of updating the attribute embedding matrix and adjacency embedding matrix is $O(k(l_a + d_a))$ and $O(k(l_x + d_x))$ respectively. Updating the eigenvectors of the attribute embedding matrix and adjacency embedding matrix requires the computational complexity $O(k^2(l_a + d_a + n))$ and $O(k^2(l_x + d_x + n))$. Finally, the computational complexity of the embedding consensus matrix is $O(k^2l)$. So the total computational complexity is $O(k^2(n + l + l_a + l_x + d_a + d_x))$.

The attribute and adjacency matrix of a substrate physical network change smoothly in continuous time steps Therefore, $\Delta L_A, \Delta L_X, \Delta D_A, \Delta D_X$ are sparse, and the number of non-zero elements, l_a, l_x, d_a, d_x, will be very small. In addition, according to spectrum method, $k \ll n$. Therefore, the computational complexity of dynamic update will be much lower than the computational complexity of static update. It is proved that RDAM is a practical dynamic update algorithm.

5.1.4 *Network Modelling*

5.1.4.1 Model Architecture

This subsection introduces the model used in the RDAM algorithm. The input of model shown in Fig. 5.3 is the embedding consensus matrix $Y^{(t)}$. And we will obtain the probability of the virtual node to each physical network node.

As shown in Fig. 5.3, the model consists of four layers. The first layer is the input layer, where $Y_i^j (1 \leqslant i \leqslant n, 1 \leqslant j \leqslant l)$ represents the value of the j-th dimension in the hidden layer representation of the i-th substrate node. The second layer is the convolutional layer [25]. After the convolution layers, we could get a vector representation of n-dimensional available physical nodes, i.e.,

$$h_i = w \cdot Y_i + b, \tag{5.28}$$

where h_i is the output of the i-th convolution layer, w is the weight vector of the convolution kernel, and b is the bias.

The third layer is the softmax layer. The output of the convolutional layer h is passed to the softmax layer. We will obtain a probability vector that represents the virtual request node selecting each substrate node. For each physical node, the probability softmax is calculated as

$$p_i = \frac{e^{h_i}}{\sum_j e^{h_j}}. \tag{5.29}$$

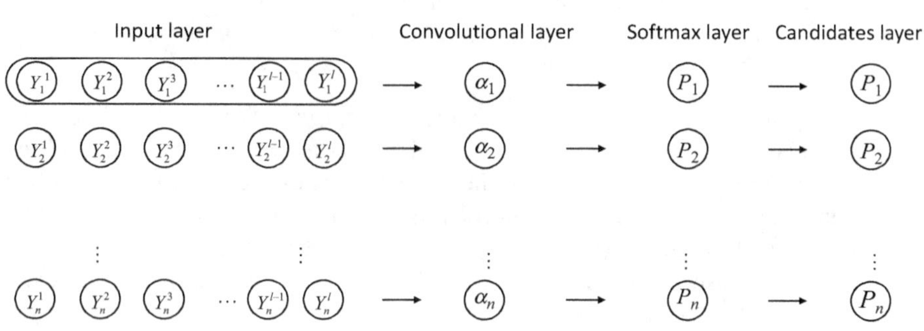

Fig. 5.3 The model architecture of RDAM

Softmax is a generalization of logistic functions. It can obtain the probability of each dimension of a high-dimensional vector. The probability of each dimension ranges (0,1), and the sum of all dimensions probability is 1.

The fourth layer is candidate node layer. Through this model, each physical node has its probability. However, in actual situation, some physical node does not satisfy the virtual request node. Therefore, the layer will have a filter to remove the unsatisfied nodes. In this case, we will select the physical node with the highest probability after candidate node layer.

5.1.4.2 Reinforcement Learning Agent

Let us consider a substrate network with n nodes and m links. At an arbitrary time, the virtual network embedding solver receives a virtual network request G^V that requires p virtual nodes and q virtual links. We define the MDP corresponding to virtual node mapping of G^V as a finite-horizon MDP M_{G^V}. The decision-making agent consecutively selects p substrate nodes for embedding nodes of G^V, yielding to p decision-making instances at discrete times t. We assume that in a given state s_t^V, the agent tries to identify a substrate node $n^V \in N^S$ for embedding the first element $n^V \in N^V$.

The state of S_t^V at the decision making instance t is defined as

$$S_t^V = (N_t^V = N_{t-1}^V \setminus \{n_{t-1}^V\}, N_t^S, N_{t-1}^S \setminus \{n_{t-1}^S\}), \tag{5.30}$$

where n_{t-1}^S is the substrate node selected for embedding the virtual node n_{t-1}^V in the previous time step. In the initial state, no virtual node has been embedded and, thus, all substrate nodes are available for embedding the first virtual node. Hence, $N_1^S = N^S$ and $N_1^V = N^V$.

The agent selects a node $n^S \in \{N_t^S \cap N^S(n_t^V)\}$ from the set of viable actions,

$$A_t^V = \{\varepsilon\} \cup \{(n_t^V, n^s) : \forall n^S \in \{N_t^S \cap N^S(n_t^V)\} \tag{5.31}$$

where ε denotes an arbitrary action that forces a transition to a new state. As the result of selecting a substrate node n_t^s for embedding the virtual node n_t^V, the agent receives a reward.

The size of the MDP state space grows exponentially with the number of state variables. Complexity of exact algorithms for solving MDP such as Q-learning is polynomial in the size of the state space. Therefore, finding exact solutions for MDPs with large number of state variables is intractable. We utilize policy gradient to solve this MDP problem and we will illustrate policy gradient method for virtual embedding problem in the following subsection.

5.1.5 *Training and Testing*

In supervised learning, samples include features and labels. After dealing with features, the model will obtain a predicted label. The distance from the real label and the predicted label is used as loss function. Supervised learning is a process continuously reducing the loss function value and making the predicted label closer to the real label. There is no real label of the samples in reinforcement learning. However, reinforcement learning [27] has an evaluation metric. After dealing with features, reinforcement learning chooses a predicted label randomly. If the predicted label yields a better evaluation metric value, it indicates that the direction of the model prediction is correct. In this case the parameters of the model are encouraged to be trained in this direction. If the evaluation metric value brought by the predicted label is small or even negative, it means that the direction of the model prediction is not correct. The direction of parameter training of the model needs to be adjusted.

In virtual network embedding process, we assume that the physical network has n nodes and each node embedding matrix represents l dimensions. The embedding matrix $(n \times l)$ of physical network is passed as input to the model. At this point, we cannot directly select the physical network node with the highest probability because the network model parameters are initialized randomly. If the node with the highest probability is selected, the model is always biased. Therefore, we need to find a balance between the exploration of a better solution and the exploitation of the existing model. We randomly select the i-th node from the probability vector $\mathbf{P}$ and construct a one-hot encoded vector. That is, the vector has n dimensions, only the i-th element is 1, and the rest are 0. Then, the loss function can be written as

$$L(y, p) = -\sum_{i} y_i log(p_i), \tag{5.32}$$

where y_i and p_i are the value of the randomly selected one-hot vector and the predicted probability vector respectively. Then we utilize the gradient derivative to train the model. When this randomly selected node can yield a large evaluation metric value, the direction of the model training is more inclined to make similar decisions. When this randomly selected node yields a small evaluation metric value, the model parameter training is not encouraged to make similar decisions. Therefore, we update the gradient by

$$g := \alpha \cdot r \cdot g \tag{5.33}$$

where α is the learning rate which controls the speed of the model training. When α is too large, the training process may not converge and the global optimal solution may be missed. When α is too small, the training process is too slow. We need to choose an appropriate learning rate. The larger rewards will have a greater impact on learning agents by multiplying rewards with gradients. In this case, the model can be more inclined to make similar decisions. And decisions that get smaller rewards or negative rewards will have a smaller impact for learning agents.

For a virtual network request, there are generally multiple virtual network request nodes. After dealing with each virtual request node, the RDAM algorithm will stack the gradient instead of directly applying in the model because the virtual network request may fail. If the embedding task fails, the corresponding stacked gradients will be cleared and the next virtual network request is processed.

After the number of virtual network requests reaches the number of batches, all the gradients in the stack are applied to the model, and then the stack is cleared. The reason for we utilize the batch gradient descent is that the gradient update requires a lot of time. If the batch gradient is adopted, it will save much time. Secondly, the batch gradient averages the gradient in the batch size and the training results are more stable.

A complete virtual network embedding training process is shown in Algorithm 2. Lines 7–10 are the process of node mapping, and lines 11–13 are the process of link mapping. Line 28 illustrates when the mapping fails, it clears the gradient in the stack and starts to train the next virtual network embedding request. Lines 21–23 update the batch-size gradients and clear the request counter.

Algorithm 2 Training process

Input: Number of epochs $numEpoch$; Learning rate α; Training set;
Output: Trained parameters in policy network;

```
1: Initialize all the parameters in policy network;
2: while iteration < numEpoch do
3:     for req ∈ trainingSet do count=0;
4:         for node ∈ req do
5:             M_f =getFeatureMatrix();
6:             p=policyNet.getOutput(M_f); //Get the probability distribution from policy network
7:             host=sample(p); //Sample from the probability distribution to choose a node as host
8:             computeGradient(host);
9:         end for
10:        if isMapped(∀ node ∈ req) then
11:            bfsLinkMap(req);
12:        end if
13:        if isMapped(∀ node ∈ req, ∀ link ∈ req) then
14:            reward=revToCost(req); //Compute revenue to cost ratio
15:            multiplyGradient(reward, α); //Compute the final gradients
16:        else
17:            clear the stacked gradients;
18:        end if
19:        ++counter;
20:        if counter reach the batch size then
21:            apply gradients to parameters;
22:            counter=0;
23:        end if
24:    end for
25:    ++iteration;
26: end while
```

Algorithm 3 Testing process

Input: Testing set;
Output: Long-term average revenue, acceptance ratio, long-term revenue to cost ratio;

```
Initialize all the parameters in policy network;
for req ∈ testSet do
    for node ∈ req do
        M_f =getFeatureMatrix();
        host=maxProbabilty(p); //Greedy strategy
    end for
    bfsLinkMap(req);
    if isMapped(∀ node ∈ req, ∀ link ∈ req) then
        signal(SUCCESS);
    end if
end for
```

The pseudocode for the test stage is shown in Algorithm 3. In the testing stage, the RDAM utilizes the greedy policy to select the substrate node from the largest probability node for mapping.

5.1.5.1 A Simple Example for RDAM Algorithm

Considering the topologies of the virtual network and the substrate network as illustrated in Fig. 5.2, we can obtain that the adjacency matrix X at initial time is

$$X = \begin{bmatrix} 0 & 1 & 0 & 1 \\ 1 & 0 & 1 & 0 \\ 0 & 1 & 0 & 1 \\ 1 & 0 & 1 & 0 \end{bmatrix}. \quad (5.34)$$

We assume that the attribute matrix A at initial time is

$$A = \begin{bmatrix} 20 & 10 & 10 & 30 \\ 10 & 10 & 10 & 20 \\ 20 & 10 & 30 & 10 \\ 10 & 20 & 10 & 30 \end{bmatrix}. \quad (5.35)$$

As we illustrated in Algorithm 1, the first step is to get the static representation of substrate physical network using spectrum method. We set the dimension k of the embedding matrix Y_X 3 and set the dimension l of the embedding matrix Y_A 2. We can obtain the consensus matrix Y

$$Y = \begin{bmatrix} 0.24 & 7.07 \\ 2.42 & 1.18 \\ 5.22 & 1.82 \\ 2.91 & 6.91 \end{bmatrix}. \tag{5.36}$$

The embedding matrix $Y \in \mathbb{R}^{n \times l}$ of physical network is passed as input to the model. Then we can train the model following the Algorithm 2. We utilize the perturbation theory defined in Algorithm 1 to capture the changes of nodes and links in the substrate network under continuous time. When the training process ends, the RDAM utilizes the greedy policy to select the substrate node from the largest probability node for mapping.

5.1.6 Experiments

5.1.6.1 Datasets

This part utilizes the GT-ITM tool to generate the substrate network topology which is commonly used in virtual network embedding algorithms. Finally, we form a substrate network with approximately 100 nodes and 500 links, which is the size of a medium-sized ISP. The CPU resources of each substrate node are uniformly distributed from 50 to 100 units, and the bandwidth resources of each substrate link are uniformly distributed from 20 to 50 units.

Similarly, we generate some virtual network requests. Each request has 2–10 virtual nodes. The CPU demand is uniformly distributed from 0 to 50. These virtual nodes are connected to each other with a probability of 0.5 forming an average of $(n-1)/4$ virtual links. Bandwidth requirements for virtual links are uniformly distributed from 0 to 50 units. The process of a virtual request is a Poisson process with an average of 4 requests over 100 time units.

In order to verify the generalization ability of the algorithm, we constructed 2000 virtual network requests. The training set contains 100 substrate physical nodes and the first 1000 virtual network requests. The test set contains 100 substrate physical nodes and the last 1000 virtual network requests. First, the model is trained by the training set. If the trained network parameters have improved in the training set, the same network parameters are used to perform virtual network embedding on the test set. In this way we can observe the generalization effect of the training model on the test set.

5.1.6.2 Experimental Settings

Four-layers network is constructed by the Tensorflow and the network is initialized using the parameters by normal distribution. Some hyperparameter settings are involved in the training process. For example, the batch size is set to 100 and the learning rate is set to 0.005. In the representation process of the physical network, the dimension k of the embedding matrix is set to 20, and the dimension l of the consensus matrix is set to 4.

5.1.6.3 Training Results

The reinforcement training process is more difficult to converge than the supervised learning process. It requires an interaction process with environment to continuously perceive the state of the environment. The learning agent makes decisions, adopts certain behavior, receives rewards from the environment, and then adjusts its strategies according to the rewards. Especially in the problem of virtual network embedding such an NP-hard, it takes a very long time to converge.

Figure 5.4 shows the change of long-term average revenue, long-term revenue to cost ratio and long-term acceptance ratio in 100 epochs. At the beginning of the training (0 <epoch< 20), all three evaluation metrics perform poorly because the parameters of the model are randomly initialized. In the middle of the training (20 <epoch< 80), the evaluation metrics values start to get better, because the random selection of physical nodes in the training process allows the model to explore all possibilities of selecting physical nodes. When randomly selected nodes have large revenues, the gradient update of the model will be large. It donates that the learning agent remembers the rewards of such decisions and produces similar decisions in subsequent decisions. At the later stage of training (80 <epoch< 100), the metric values begin to fluctuate around a certain value, because the model is still exploring the possibility of physical node selection, but the training process at this time has converged.

Figure 5.5 shows the decreasing trend of the cross-entropy loss function during training. It can be seen that the value of the loss function is constantly decreasing, which also proves the effectiveness of the model. In the last 20 epochs, the loss function tends to be a constant value, which means that the training process has converged.

5.1.6.4 Testing Results

In order to verify the generalization of the model, we conduct experiments on the test set. The other three algorithm were selected. The first one is the baseline algorithm [37]:

$$H(n^s) = CPU(n^S) \sum_{l^S \in L(n^s)} BW(L^S) \tag{5.37}$$

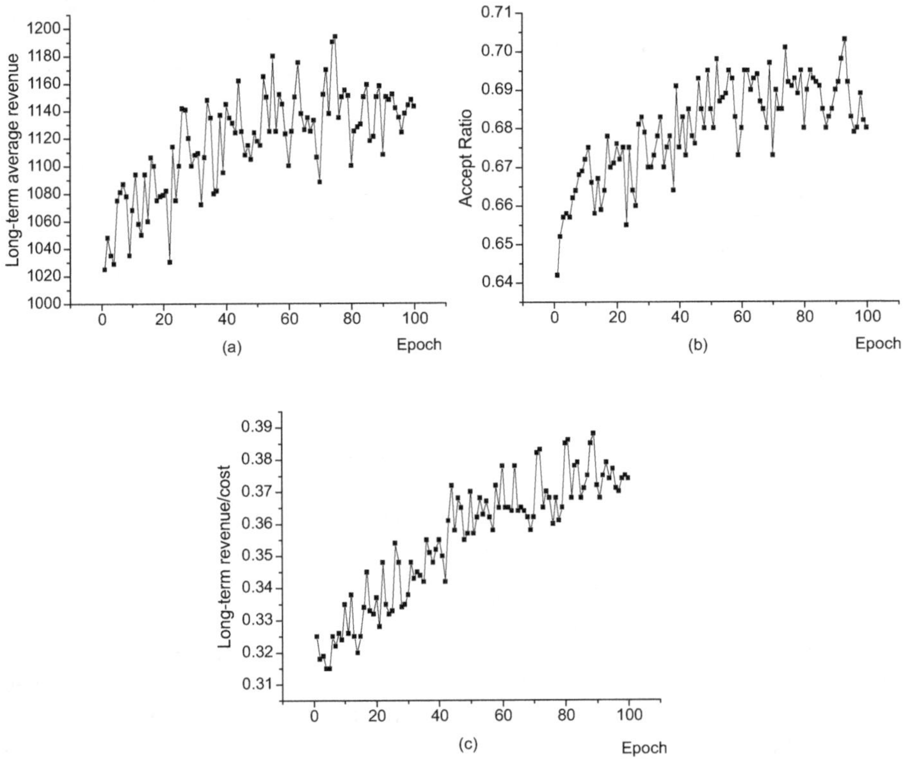

Fig. 5.4 Performance on training set. (**a**) The change of long-term average revenue in 100 epochs. (**b**) The change of long-term revenue to cost ratio in 100 epochs. (**c**) The change of long-term acceptance ratio in 100 epochs

In the training process, the long-term revenue consumption ratio is used as an evaluation metric. In each epoch, if the long-term average revenue is higher than the current best long-term average revenue, then the best long-term average revenue will update. In addition, if the current best long-tern average revenue updates, the model with current parameter will be used to perform virtual network embedding algorithm on the test set. Obviously, the best evaluation metric value will be updated frequently and the last result is the global best result.

In the test phase, the start time of the virtual network request is 22 and the end time is 30,000. Assuming the time unit is 1000, Fig. 5.6 shows the change of the evaluation metric value in thirty time units. In Fig. 5.6a and b, the long-term average revenue and acceptance ratio will decrease in the initial stage, because physical resources will be gradually consumed as virtual network requests arriving. The initial period of long-term revenue and consumption ratio will not be significantly reduced, because it has nothing to do with the number of physical resources.

It can be observed that the RL algorithm has a higher accept ratio at the very beginning. It may be because the RL algorithm allocates the top resources. As the

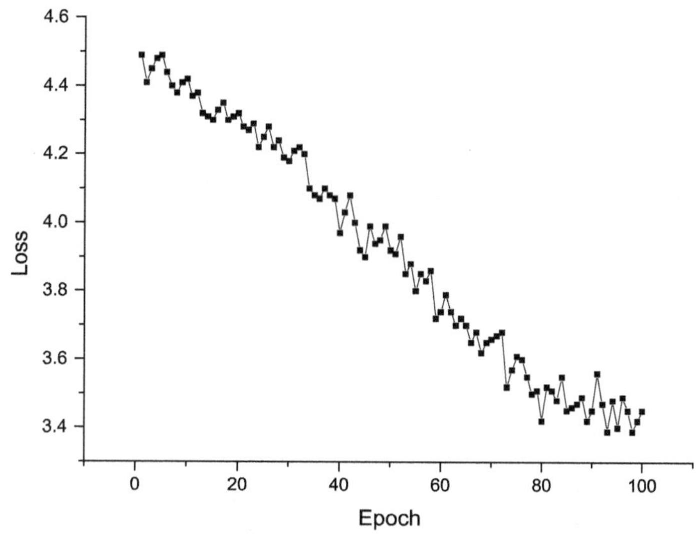

Fig. 5.5 Loss on training set

three metrics tend to be stable, it can be observed that in the three evaluation metrics, the convergence trend and convergence value of the RDAM algorithm are better than the other three algorithms. We can conclude that the reinforcement learning agent learns the relationship of the physical network nodes during the training stage, and the model can be generalized during the testing stage.

5.2 Virtual Network Embedding Based on Policy Network and Reinforcement Learning

The combination of network virtualization and software defined networks is considered as the foundation towards the next generation of Internet architecture [28, 30]. Network virtualization enables the coexistence of multiple heterogeneous virtual networks on a shared network [19, 31–33, 35]. For Internet Service Providers (ISPs), it enables new business models of hosting multiple concurrent network services on their infrastructures. Decisions for embedding are challenging problems for ISPs since it determines the effectiveness of utilization of network resources [16]. A sub-optimal embedding algorithm will decrease the overall capacity of the infrastructure and lead to cost of revenue for ISPs. A virtual network consists of several virtual nodes (e.g. virtual routers), connected by a set of virtual links. The purpose of virtual network embedding is to map virtual networks to a shared physical network while providing the requests with adequate computing and bandwidth resources.

However, the virtual network embedding problem has been proved to be NP-hard [36]. As a result, a large number of heuristic algorithms have been proposed

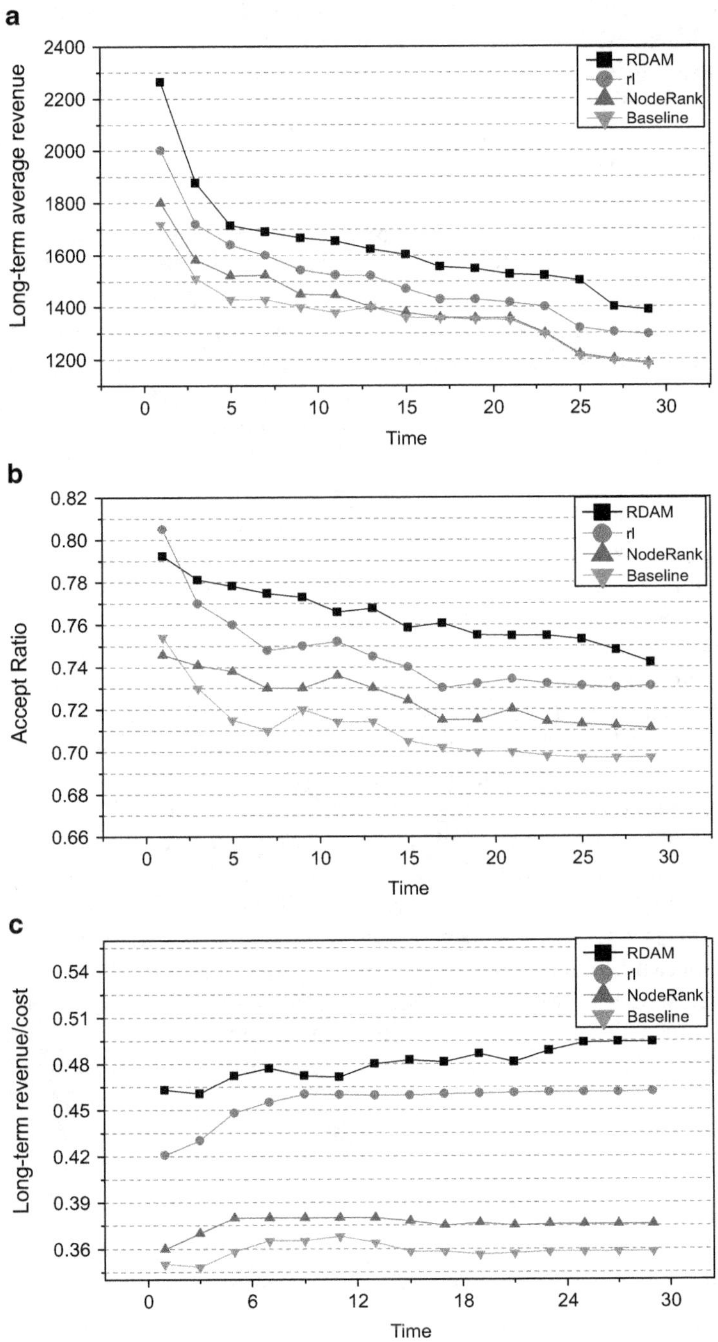

Fig. 5.6 Performance on testing set. (**a**) The change of long-term average revenue in 30 time units. (**b**) The change of long-term revenue to cost ratio in 30 time units. (**c**) The change of long-term acceptance ratio in 30 time units

[8, 37–39, 42], but most of them rely on artificial rules to rank nodes or make mapping decisions. The parameters in these algorithms are always fixed and cannot be optimized, making the embedding decisions sub-optimally. On the other hand, in prior works, the information about substrate network and the knowledge about virtual network embedding hidden in historical network request data have always been overlooked. Historical network requests are a good representation of temporal distribution and resource demands in the future.

In recent years, big data, machine learning and artificial intelligence have exciting breakthroughs achieving state of the art results such as natural language understanding and object detection [17]. Machine learning algorithms process a large amount of data collected during a period and automatically learn the statistical information from the data to give classification or prediction. Reinforcement learning, as a widely-used technique in machine learning, has shown a great potential in dealing with complex tasks, e.g., game of go [43], or complicated control tasks such as auto-driving and video games [29, 45]. The goal of a reinforcement learning system (or an agent) is to learn better policies for sequential decision making problems with an optimal cumulative future reward signal [44, 46].

In this part, we introduce reinforcement learning into the problem of virtual network embedding to optimize the node mapping process. Similar to earlier works [8, 37, 48], our work is based on the assumption that all network requests follow an invariable distribution. We divide our network request data into a training set and a testing set, to train our Reinforcement Learning Agent (RLA) and evaluate its performance respectively. We devise an artificial neural network called policy network as the RLA, which observes the status of substrate network and outputs node mapping results. We train the policy network with historical network request data using policy gradient through back propagation. An exploration strategy is applied in the training stage to find better solutions, and a greedy strategy is applied in evaluation to fully evaluate the effectiveness of the RLA. Extensive simulations show that the RLA is able to extract knowledge from historical data and generalize it to incoming requests. To the best of our knowledge, this work is the first to utilize historical network requests data and policy network based reinforcement learning to optimize virtual network embedding automatically. The RLA outperforms two representative embedding algorithms based on node ranking in terms of long-term average revenue and acceptance ratio, while making a better utilization of network resources [41].

5.2.1 Network Modelling

In this subsection, we present a network model and formulate the virtual network embedding problem with description of its components. The notations used in this section are shown in Table 5.2.

Table 5.2 Frequently used notations

G^S	Substrate network
N^S	Nodes of substrate network
L^S	Links of substrate network
A_N^S	Node attribute of substrate network
A_L^S	Link attribute of substrate network
G^V	Virtual network of a certain virtual request
N^V	Nodes of a virtual network
L^V	Links of a virtual network
A_N^V	Constraints of substrate nodes
A_L^V	Constraints of substrate nodes

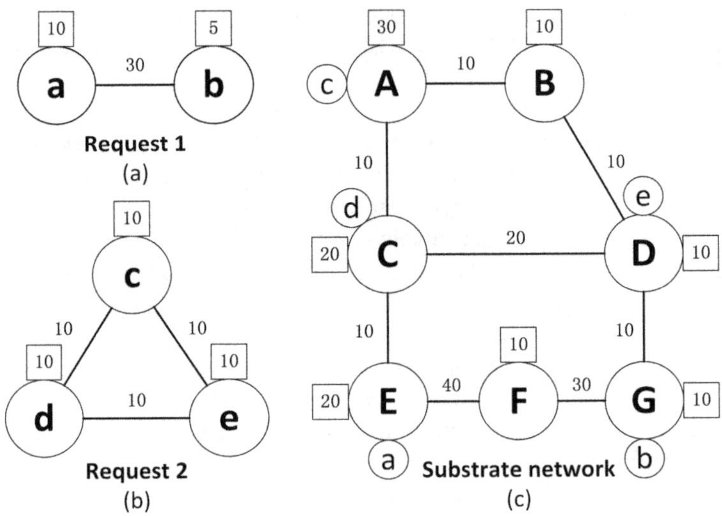

Fig. 5.7 An example of virtual network embedding

Figure 5.7 shows the mapping process of two different virtual network requests. A substrate network is represented as an undirected graph $G^S = (N^S, L^S, A_N^S, A_L^S)$, where N^S denotes the set of all the substrate nodes, L^S denotes the set of all the substrate links, A_N^S and A_L^S stand for the attributes of substrate nodes and links respectively. In consistency with earlier works[8, 37], in this part we consider computing capability as node attribute and bandwidth capacity as link attribute. Let P^S denote the set of all the loop-free paths in substrate network. Figure 5.7c shows an example of a substrate network, where a circle denotes a substrate node, and a line connecting two circles denotes a substrate link. The number in a square box denotes the CPU(computing) capacity of that node, and the number next to a substrate link denotes the bandwidth of that link.

Similarly, we also use an undirected graph $G^V = (N^V, L^V, C_N^V, C_L^V)$ to describe a virtual network request, where N^V denotes the set of all the virtual nodes in the request, L^V denotes the set of all the virtual links in the request, C_N^V and C_L^V stand for the constrains of virtual nodes and links respectively. To map a virtual node to a substrate node, the computing capacity of the substrate node must be higher than that is required by the virtual node. To map a virtual link to a set of substrate links, the bandwidth of each substrate link must be higher than that is required by the virtual link. Figure 5.7a and b show two different virtual requests. Additionally, we use t to denote the arrival time of a virtual request, and use t_d to denote the duration of the virtual request.

When a virtual request arrives, the objective is to find a solution to allocate different kinds of resources in the substrate network to the request while satisfying the requirements of the request. If such a solution exists, then the mapping process will be executed, and the request will be accepted. Otherwise the request will be rejected or delayed. The virtual network embedding process can be formulated as a mapping M from G^V to $G^S : G^V(N^V, L^V) \rightarrow G^S(N', P')$, where $N' \subset N^S$, $P' \subset P^S$.

The main goal of virtual network embedding is to accept as many requests as possible to achieve maximum revenue for an ISP, when the arrival of virtual network requests follows an unknown distribution of time and unknown resource requirements [47]. Consequently, the embedding algorithm must produce efficient mapping decisions within an acceptable period. As shown in Fig. 5.7, virtual nodes a and b in request 1 are mapped to substrate nodes E and G respectively, and virtual nodes c, d and e in request 2 are mapped to substrate nodes A, C and D respectively. Note that the embedding result of request 1 is not optimal. For example, the cost of bandwidth in the substrate network can be significantly reduced by moving a to F.

To determine the performance of embedding algorithms, most works use certain metrics such as a long-term average revenue, a long-term acceptance ratio, and a long-term revenue to cost ratio. The revenue measures the profit of an ISP for accepting a certain virtual request, and it depends on the amount of requested resources and the duration of it. Similar to the earlier works presented in [8, 37], we define the revenue of accepting a virtual network request as follows:

$$R(G^v, t, t_d) = t_d \cdot [\sum_{n^V \in N^V} CPU(n^V) + \sum_{l^V \in N^V} BW(l^V)] \tag{5.38}$$

where $CPU(n^V)$ and $BW(l^V)$ denote the computing resource that a virtual node n^V requires and the bandwidth resource that a virtual link l^V requires respectively. As shown in the formula, virtual requests having more resources requirements or lasting longer have more revenue.

The cost function measures the efficiency of utilizing substrate network resources. We define the cost of accepting a virtual request as follows:

$$C(G^v, t, t_d) = t_d \cdot [\sum_{l^V \in N^V} \sum_{l^V \in N^V} BW(l^V)] \tag{5.39}$$

where $P(l^V)'$ denotes the set of substrate links where virtual link l^V is embedded. $C(G^V, t, t_d)$ computes the actual consumption of bandwidth resource for embedding request G^V. When accepting a virtual request, the CPU consumption is always fixed, but the bandwidth consumption may vary depending on the performance of embedding algorithm discussed above.

Following the works presented in [8, 37, 49], we use a long-term average revenue to evaluate the overall performance of our embedding method defined as:

$$\lim_{T\to\infty} \frac{\sum_{t=0}^{T} R(G^v, t, t_d)}{T} \tag{5.40}$$

where T is the time elapsed. A higher long-term average revenue leads to a higher profit for the ISP. Another important metric to evaluate the mapping algorithm is a long-term acceptance ratio, which means the ratio of accepted requests to the total number of requests arrived. A higher long-term acceptance ratio means the proposed algorithm manages to serve more virtual requests.

Finally, a better utilization of substrate network resources would lead to a high long-term average revenue with comparatively low cost of substrate network. The long-term revenue to cost ratio, defined as follows, measures the utilization of substrate network resources:

$$\lim_{T\to\infty} \frac{\sum_{t=0}^{T} R(G^v, t, t_d)}{\sum_{t=0}^{T} C(G^v, t, t_d)} \tag{5.41}$$

A higher long-term revenue to cost ratio shows that the proposed algorithm is able to generate more profit with a comparatively less cost to network resources.

We will use these metrics mentioned above to evaluate the performance of our embedding method in the following subsections.

5.2.2 *Embedding Algorithm*

In this subsection, we present the details of the proposed policy network based reinforcement learning algorithm. Specifically, we apply the reinforcement learning agent in the node mapping stage to derive the probabilities of choosing nodes. The agent takes a feature matrix extracted from the substrate network as input, and makes decisions based on a policy network which is trained from historical data.

5.2.2.1 Feature Extraction

Every substrate node has several attributes, such as CPU capacity and the total amount of bandwidth of the adjacent links [51]. A thorough knowledge of substrate network is crucial for the reinforcement learning agent to establish a basic understanding of its state and generate efficient mapping [52, 56]. To facilitate the agent to choose the substrate nodes, we need to extract features of each substrate node and use them as input to the policy network.

We extract four features for each substrate node listed as follows:

- Computing capacity (CPU): The CPU capacity of a substrate node n^S has a large impact on its availability. The substrate nodes with a higher computing capacity are likely to host more virtual nodes.
- Degree (DEG): The degree of a substrate node n^S indicates the number of links connected to it. A substrate node with more adjacent links is more likely to find paths to other substrate nodes.
- Sum of bandwidth ($SUM^{(BW)}$): Every substrate node is connected to a set of links. A substrate node n^S has a sum of bandwidth resources of its neighboring links:

$$SUM^{(BW)}(n^S) = \sum_{l^S \in L(n^S)} BW(l^S) \tag{5.42}$$

  where $L(n^S)$ is the neighboring links of n^S and $BW(l^S)$ is the bandwidth resource of a substrate link l^S. When a substrate node has access to more bandwidth, mapping a virtual node to it may lead to better link mapping options.
- Average distance to other host nodes $AVG^{(DST)}$: When mapping a virtual node, we also take into consideration the positions where other virtual nodes in the same request are mapped. By choosing a substrate node close to those already mapped, the cost of substrate link bandwidth can be reduced. We measure the distance between two substrate nodes in terms of the number of links along the shortest path. The shortest path is computed following the Floyd—Warshall algorithm [53]. We take an average of the distance from a substrate node n^S to another host nodes $\widetilde{N}^S$ for the same request:

$$AVG^{(DST)}(n^S) = \frac{\sum_{\widetilde{n}^S \in \widetilde{N}^S} DST(n^S, \widetilde{n}^S)}{|\widetilde{N}^S| + 1} \tag{5.43}$$

  where $DST(n^S, \widetilde{n}^S)$ is the distance from node n^s to node $\widetilde{n}^S$.

In fact, the features that we can extract from the substrate nodes would be far more than listed above. More features would bring more information about the substrate network which leads to a better performance of the learning agent. It should be noted that extracting more features from the substrate network adds complexity in computation.

After extracting the features of the kth substrate node n_k^S, we take their normalized values and concatenate them into a feature vector v_k:

$$v_k = (CPU(n_k^S), DEG(n_k^S), SUM^{(BW)}(n_k^S), AVG^{(DST)}(n_k^S))^T \tag{5.44}$$

The purpose of normalization is to accelerate the training process and enable the agent to converge quickly. We concatenate all feature vectors of substrate nodes to produce a feature matrix M_f where each row is a feature vector of a certain substrate node:

$$M_f = (v_1, v_2 \cdots v_{|N^S|})^T \tag{5.45}$$

The feature matrix serves as an input to the learning agent. The feature matrix is updated along with the changing substrate network from time to time.

5.2.2.2 Policy Network

In this work, we implemented an artificial neural network called policy network as the learning agent. It takes the feature matrix as input and outputs the probabilities of mapping virtual nodes to substrate nodes.

For simplicity, we build a simple policy network with basic elements of an artificial neural network as shown in Fig. 5.8. The policy network contains an input layer, a convolutional layer, a softmax layer and finally a node filter. For each virtual node that requires a mapping, we use the policy network to choose a substrate node for it.

At the input layer, we compute the feature matrix and deliver it to the policy network. The policy network then passes the input feature matrix into a convolutional layer with one convolution kernel, where the policy network evaluates the resources of each substrate node. The convolutional layer performs a convolution operation on the input to produce a vector representing the available resources of each node:

$$h_k^c = \omega \cdot v_k + b \tag{5.46}$$

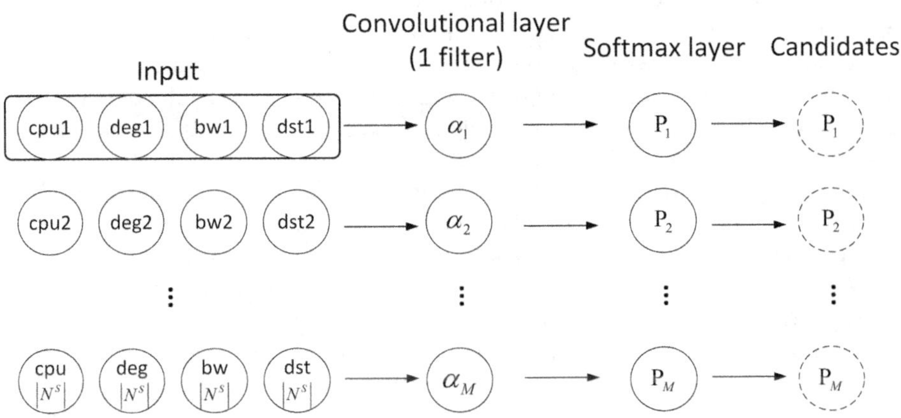

Fig. 5.8 Policy network

where h_k^c is the kth output of the convolutional layer, ω is the convolution kernel weight vector, and b is bias.

Then the vector is transmitted to a softmax layer to produce a probability for each node which indicates the likelihood of yielding a better result if mapping a virtual node to it. For the kth node, the probability p_k is computed as:

$$p_k = \frac{e^{h_k^c}}{\sum_i e^{h_i^c}} \tag{5.47}$$

The softmax function is a generalization of the logistic regression. It turns the n-dimensional vector into real values between 0 and 1 that add up to 1. The output of the softmax function indicates a probability distribution over n different possible mappings. Some of the nodes are not able to host the virtual node in concern because they do not have enough computing resources. We add a filter to choose a set of candidate nodes with enough CPU capacities.

5.2.2.3 Training and Testing

We first randomly initialize the parameters in the policy network, and train it for several epochs. For every virtual node in each iteration, a feature matrix is extracted from the substrate network which serve as input to the policy network. The policy network outputs a set of available substrate nodes as well as a probability for each node. The probability of each node represents the likelihood that mapping a virtual node to it will yield a better result. In the training stage, we cannot simply select the node with a maximal probability as the host because that the model is randomly initialized, which means the output could be biased and better solutions might exist. In other words, we need to reach a balance between the exploration of better solutions and the exploitation of current model. To this end, we generate a sample from the set of available substrate nodes according to their probability distribution that the policy network outputs, and select a node as the host. We repeat this process until all the virtual nodes in a virtual request are assigned and proceed to link mapping. If no substrate node is available, the mapping fails due to a lack of resources. For link mapping, we apply a breadth-first search to find the shortest paths between each pair of nodes.

In supervised learning, each piece of data in the training set corresponds to a label indicating the desired output of the model. With each output from model and the corresponding label, a loss value is computed which measures the deviation between them. The loss value for each piece of data in the training set sums up to an aggregated loss value, and the training stage aims to minimize the aggregated loss value. However, in reinforcement learning tasks such as virtual network embedding, data in the training set does not have corresponding labels. The learning agent relies on reward signals to know if it is working properly. A big reward signal informs the learning agent that its current action is effective and should be continued. A small reward signal or even a negative reward signal shows that the current action is erroneous and should be adjusted. The choice of reward is critical in reinforcement learning as it directly influences the training process and determines the final policy. Here, we use the revenue to cost ratio of a single virtual request as the reward for every virtual node in this request because this metric represents the utilization efficiency of the substrate resources. Then we apply policy gradient method to train the policy network.

The actual implementation of the proposed algorithm is non-trivial since we cannot provide each output with a label. As a result, we temporarily consider every decision that the agent makes to be correct by introducing a *hand-crafted label* into our policy network. Assume that we choose the ith node, then the *hand-crafted label* in policy network would be a vector y filled with zeros except the ith position which is one. Then we calculate the cross-entropy loss:

$$L(\mathbf{y}, \mathbf{p}) = -\sum_{i} y_i log(p_i) \tag{5.48}$$

where y_i and p_i are the ith element of *hand-crafted label* and the output of policy network respectively. We use backpropagation to compute the gradients of parameters in the policy network. Since we use *hand-crafted label*, we stack the gradients g_f rather than applying them immediately. If our algorithm fails to embed a virtual request, the corresponding stacked gradients will be aborted since we cannot determine the reward signal. If a virtual request has been successfully mapped, we compute its revenue to cost ratio as a reward r. Then we multiply the stacked gradients by using the reward and an adjustable learning rate α to achieve the final gradients:

$$g = \alpha \cdot r \cdot g_f \tag{5.49}$$

The learning rate α is introduced to control the magnitude of gradients and the computation speed of training. If the gradients are too large, the model becomes unstable and may not improve through the training process. On the other hand, too small gradients make training extremely slow. Therefore the learning rate needs to be tuned carefully. It can be observed from Eq. (5.49) larger rewards make the corresponding gradients more significant than small ones. As a result, the choices that lead to larger rewards have larger impact on the learning agent, making it more prone to make similar decisions. When we stack a batch of gradients, we apply them to parameters and update the policy network. There are two reasons for batch updating—one is that parameter updating normally takes a long time, but doing that in batches speeds up this process. Another reason is that batch updating averages over the gradients and is more stable. The training process is shown in Algorithm 4. Lines 7–10 show node mapping stage where we compute the gradients in line 10, lines 11–13 show the link mapping stage.

Algorithm 4 Training process

Input: Number of epochs, *numEpoch*; Learning rate, α; Training set;
Output: Trained parameters in policy network;

```
Initialize all the parameters in policy network;
while iteration < numEpoch do
    for req ∈ trainingSet do
        counter=0;
        for node ∈ req do
            M_f=getFeatureMatrix();
            p=policyNet.getOutput(M_f) //Get the probability distribution from policy network;
            host=sample(p) //Sample from the probability distribution to choose a node as host;
            computeGradient(host);
        end for
        if isMapped(∀ node ∈ req) then
            bfsLinkMap(req);
        end if
        if isMapped(∀ node ∈ req, ∀ link ∈ req) then
            reward=revToCost(req) //Compute revenue to cost ratio;
            multiplyGradient(reward, α) //Compute the final gradients;
        else
            clear the stacked gradients;
        end if
        ++counter;
        if counter reach the batch size then
            apply gradients to parameters;
            counter=0;
        end if
    end for
    ++iteration;
end while
```

In the testing stage, we apply a greedy strategy where we directly choose the node with the highest probability as the host. The testing algorithm is shown in Algorithm 5.

Algorithm 5 Testing process

Input: testing set;
Output: long-term average revenue, acceptance ratio, long-term revenue to cost ratio;

```
Initialize all the parameters in policy network;
for req ∈ testSet do
    for node ∈ req do
        M_f=getFeatureMatrix();
        host=maxProbablity(p) //Greedy strategy;
    end for
    bfsLinkMap(req);
    if isMapped(∀ node ∈ req, ∀ link ∈ req) then
        signal(SUCCESS);
    end if
end for
```

5.2.3 Evaluation

We conducted a number of simulation tests to evaluate the performance of the proposed reinforcement learning algorithm and compared with other embedding algorithms.

We employed GT-ITM tool [54] to generate a substrate network with 100 nodes and approximately 550 links, which is about a middle-sized ISP. The computing capacity of every substrate node is a real number that follows a uniform distribution between 50 and 100, and the bandwidth of every link is a real number that follows a uniform distribution between 20 and 50, which is similar to parameters presented in related work [8, 37].

We also generated a number of virtual requests, each with 2 to 10 virtual nodes. The computing capacity requirement of every virtual node followed a uniform distribution between 0 and 50 units. The bandwidth requirement of every virtual link follows a uniform distribution between 0 and 50 units. Virtual nodes were connected with a probability of 0.5 forming an average of $\frac{n(n-1)}{4}$ virtual links, where n is the number of nodes. The virtual requests arrived following a Poisson distribution with an average of 4 requests over 100 time units. The duration of every requests followed an exponential distribution with an average of 1000 time units. We generate a timeline that lasted around 50,000 time units containing about 2000 requests. We divided the requests equally into two sets—training set and testing set.

We employed TensorFlow [55] to build the policy network. We first assembled the layers mentioned in Sect. 5.2.2 and followed a normal distribution to initialize their parameters. Then we defined a tensor for gradient of every training step using the *compute_gradients* method of stochastic gradient descent [57] optimizer. When doing batch updates, we added all the gradients for each parameter and multiply them by a reward, and called the *apply_gradients* method of SGD optimizer to update all parameters in policy network. We train our agent for 100 epochs using gradient descent with a learning rate of 0.005. The batch size was set to 100 which means the parameters are updated every 100 requests.

5.2.3.1 Effectiveness of Reinforcement Learning

Compared to supervised learning, reinforcement learning has proven to be hard to train. It may take a very long time for the learning agent to stabilize, especially for a complicated problem such as virtual network embedding. We run the learning agent on the training data set for 100 epochs and observe its performance. Figure 5.9a shows the change of a long-term average revenue, the acceptance ratio and a long-term revenue to cost ratio during the training process. In the very beginning, the learning agent performs poorly because all the parameters in the policy network are initialized randomly. As the training goes, the random sampling allows the learning

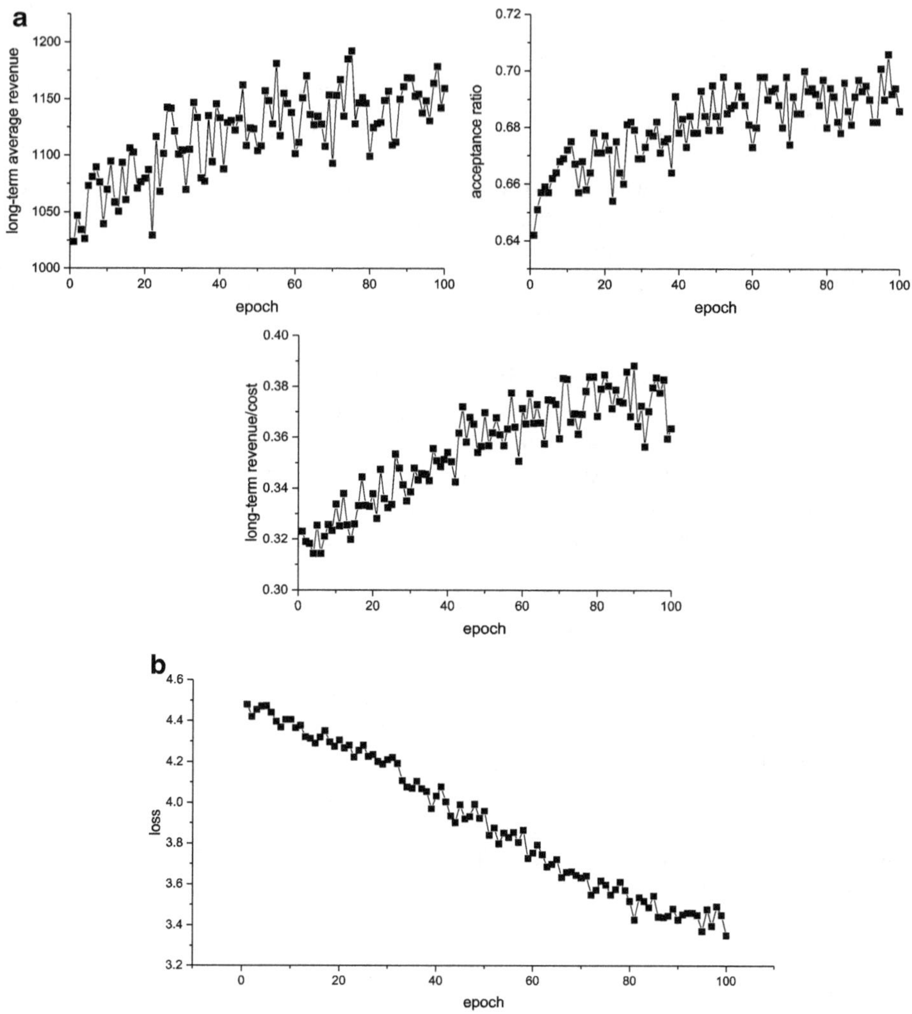

Fig. 5.9 Training process. (**a**) Performance on training set. (**b**) Loss on training set

agent to explore different possibilities. The learning agent may find a good solution occasionally and receive a great reward which helps the policy network to learn to make better decisions. Consequently, the performance starts to get better, proving the effectiveness of reinforcement learning on the task. The exploration strategy sometimes leads our agent into bad choices causing a fluctuation in its performance as the training proceeds. But such cases will lead to small rewards and have small impact on the learning agent. In the later stage of training process, the performance stopped improving due to the limited capacity of functional complexity that the policy network can handle. Eventually the learning agent reaches a certain point, and the performance stabilizes in a range. Figure 5.9b shows the cross-entropy loss during the training process. Clearly, the loss decreases through the training stage and eventually starts to stabilize in the last 10 epochs.

The result shows that the proposed reinforcement learning based algorithm is getting better performance as the training goes, which means the learning agent can adapt itself to training data.

We have proved that the reinforcement learning method can improve the performance of the learning agent on training set. But it is still unclear if the learning agent actually learns how to optimize node mapping, or simply adjusts to existing data. In order to test the generalization ability of the learning agent, we separated a testing data set that consisted of different requests from the training set and run the learning agent on it.

Different from the training process, we run the learning agent without a random sampling and applied a greedy strategy to choose a node with the maximal probability. The performance over time on the testing data set is shown in Fig. 5.10. We compare the learning agent with another two rule-based node ranking algorithms. The first one is a baseline algorithm proposed in [37] using equation:

$$H(n^S) = CPU(n^S) \sum_{l^S \in L(n^S)} BW(L^S) \tag{5.50}$$

to rank substrate nodes, where $H(n^s)$ measures the availability of substrate node n^s. The other is proposed in [8] using NodeRank algorithm to measure the importance of nodes. All the three methods followed the same breadth-first search link mapping algorithm. We measured the performance of these methods with three metrics mentioned in Sect. 5.2.1—a long-term average revenue, an acceptance ratio and a long-term revenue to cost ratio.

At the beginning, the performance of all the three algorithms on a long-term average revenue and an acceptance ratio decrease because the amount of resources of the substrate network decreases as more requests arrive. The long-term revenue to cost ratio is stable because it is irrelevant to the amount of available resources. Then the performance of all algorithms on all metrics starts to stabilize because the resources of the substrate network is depleted. The results in Fig. 5.10 show that the learning agent outperforms the other two algorithms in all the three metrics.

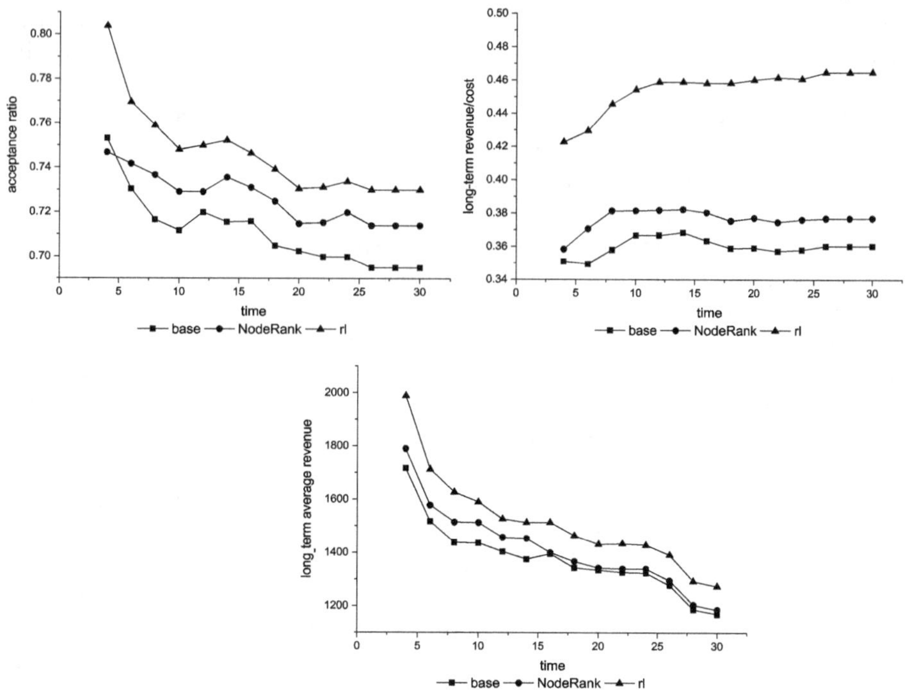

Fig. 5.10 Performance on testing set

The training data set and testing data set consist of different requests, but the learning agent is able to perform well on both data sets. The conclusion is that the learning agent does not simply adjust itself to the training set, rather it is actually capable of generalizing from the training process to acquire knowledge about the substrate network and node mapping. Note that the performance improved evidently compared to the result on the training data set, because the exploration in training process may lead to bad embedding results.

5.2.3.2 Stress Tests

We further expanded our experiments by increasing resource demands to examine the performance of the proposed algorithm under more stress. By doing this we also expected to find the circumstances where our algorithm is robust.

Figure 5.11a shows the performances of the three algorithms when we doubled the CPU requirements of the virtual nodes. Figure 5.11b shows the results when the bandwidth of virtual links were uniformly distributed between 25 and 50 units. As shown in Fig. 5.11, the proposed reinforcement learning based algorithm performs better in the later circumstance, and outperforms the other two methods against the three metrics. In the computing-intensive environment however, the

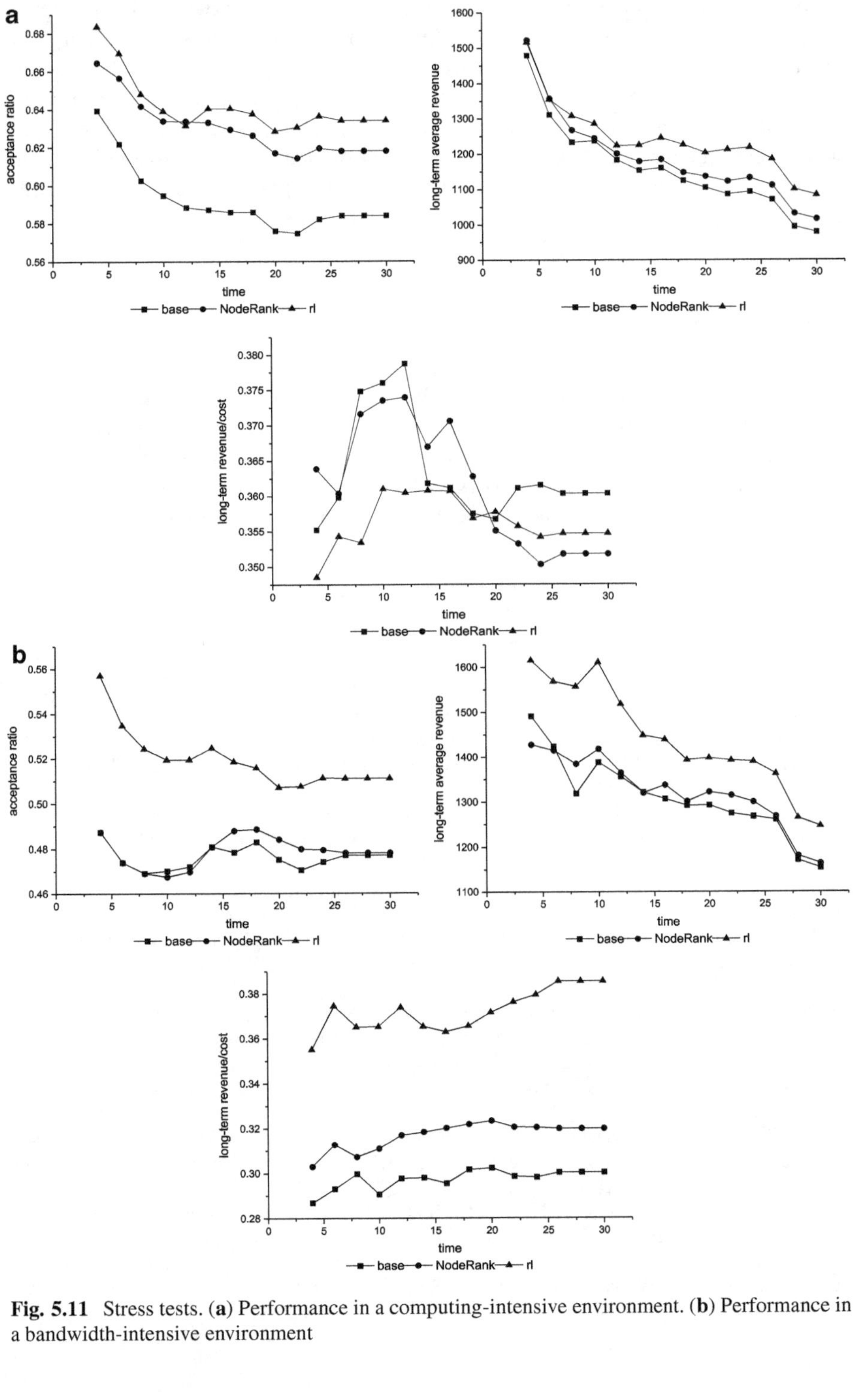

Fig. 5.11 Stress tests. (**a**) Performance in a computing-intensive environment. (**b**) Performance in a bandwidth-intensive environment

proposed algorithm achieves similar performance to the other two methods in terms of a long-term revenue to cost ratio while getting better results in the long-term average revenue and acceptance ratio. The proposed reinforcement learning based algorithm works in node mapping phase, but larger CPU requirements means less nodes with enough computing resource to choose from, which leads to relatively worse performance in computing-intensive environment. The conclusion is that the proposed algorithm can achieve comparatively better performance for bandwidth-intensive requests rather than computing-intensive ones.

5.3 Summary

In this chapter, we discuss the main challenge of intelligent network resource management and introduced several reinforcement learning algorithms. we first propose a reinforcement learning based dynamic attribute matrix representation (RDAM) algorithm for virtual network embedding. Then, we design and implement a policy network based on reinforcement learning to make node mapping decisions.

References

1. J. L. Chen, Y. W. Ma, H. Y. Kuo, and C. S. Yang, "Software-defined network virtualization platform for enterprise network resource management," *IEEE Transactions on Emerging Topics in Computing*, vol. 4, no. 2, pp. 179–186, 2016.
2. J. Lu and J. Turner, "Efficient mapping of virtual networks onto a shared substrate," *Washington University in St Louis*, 2006.
3. Z. Zhen, T. Jiang, W. Zhang, H. Yao, and S. Xiao, "Analyzing speech of patients with vocal polyps based on channel parameters and fuzzy logic systems," *Computers & Mathematics with Applications*, vol. 62, no. 7, pp. 2834–2842, 2011.
4. D. G. Andersen, "Theoretical approaches to node assignment," *Computer Science Department*, 2002.
5. Zhao, Chenglin, Haipeng, Zhou, and Zheng, "Optimization of multiband spectrum sensing for cognitive radio networks under sensing capability constrains," *China Communications*, vol. 7, no. 5, pp. 129–136, 2010.
6. C. Jiang, C. Yan, K. J. R. Liu, and R. Yong, "Network economics in cognitive networks," *IEEE Communications Magazine*, vol. 53, no. 5, pp. 75–81, 2015.
7. H. Li, M. Dong, K. Ota, and M. Guo, "Pricing and repurchasing for big data processing in multi-clouds," *IEEE Transactions on Emerging Topics in Computing*, vol. 4, no. 2, pp. 266–277, 2016.
8. X. Cheng, S. Su, Z. Zhang, H. Wang, F. Yang, Y. Luo, and J. Wang, "Virtual network embedding through topology-aware node ranking," *Acm Sigcomm Computer Communication Review*, vol. 41, no. 2, pp. 38–47, 2011.
9. Z. Tong, C. Yan, C. Jiang, and K. J. R. Liu, "Pricing game for time mute in femto-macro co-existent networks," *IEEE Transactions on Wireless Communications*, vol. 14, no. 4, pp. 2118–2130, 2015.
10. C. Jiang, X. Wang, W. Jian, H. H. Chen, and R. Yong, "Security in space information networks," *IEEE Communications Magazine*, vol. 53, no. 8, pp. 82–88, 2015.

11. C. Jiang, C. Yan, and K. J. R. Liu, "Data-driven optimal throughput analysis for route selection in cognitive vehicular networks," *IEEE Journal on Selected Areas in Communications*, vol. 32, no. 11, pp. 2149–2162, 2014.
12. C. Jiang, C. Yan, G. Yang, and K. J. R. Liu, "Indian buffet game with negative network externality and non-bayesian social learning," *IEEE Transactions on Systems Man & Cybernetics Systems*, vol. 45, no. 4, pp. 609–623, 2013.
13. C. Jiang, Y. Chen, Q. Wang, and K. J. R. Liu, "Data-driven auction mechanism design in iaaS cloud computing," *IEEE Transactions on Services Computing*, vol. 11, no. 5, pp. 743–756, 2018.
14. C. Jiang, Y. Chen, Y. H. Yang, C. Y. Wang, and K. J. R. Liu, "Dynamic chinese restaurant game: Theory and application to cognitive radio networks," *IEEE Transactions on Wireless Communications*, vol. 13, no. 4, pp. 1960–1973, 2014.
15. C. Jiang, Y. Chen, Y. Gao, and K. J. R. Liu, "Indian buffet game with negative network externality and non-bayesian social learning," *IEEE Transactions on Systems Man & Cybernetics Systems*, vol. 45, no. 4, pp. 609–623, 2015.
16. L. Feng, C. Jiang, J. Du, Y. Jian, R. Yong, Y. Shui, and M. Guizani, "A distributed gateway selection algorithm for uav networks," *IEEE Transactions on Emerging Topics in Computing*, vol. 3, no. 1, pp. 22–33, 2017.
17. Y. Kawamoto, H. Takagi, H. Nishiyama, and N. Kato, "Efficient resource allocation utilizing q-learning in multiple ua communications," *IEEE Transactions on Network Science & Engineering*, vol. PP, no. 99, pp. 1–1.
18. X. Lei, C. Jiang, C. Yan, R. Yong, and K. J. R. Liu, "Privacy or utility in data collection? a contract theoretic approach," *IEEE Journal of Selected Topics in Signal Processing*, vol. 9, no. 7, pp. 1256–1269, 2015.
19. L. Zhang, H. Yao, H. Liu, and Z. Zhou, "A novel ultra-wide band signal generation scheme based on carrier interference and dynamics suppression," *Eurasip Journal on Wireless Communications & Networking*, vol. 2010, no. 1, p. 10, 2010.
20. H. Yao, X. Chen, M. Li, P. Zhang, and L. Wang, "A novel reinforcement learning algorithm for virtual network embedding," *Neurocomputing*, vol. 284, pp. 1–9, 2018.
21. M. Belkin and P. Niyogi, "Laplacian eigenmaps and spectral techniques for embedding and clustering," *Advances in Neural Information Processing Systems*, vol. 14, no. 6, pp. 585–591, 2009.
22. U. V. Luxburg, "A tutorial on spectral clustering," *Statistics & Computing*, vol. 17, no. 4, pp. 395–416, 2007.
23. G. Peters and J. H. Wilkinson, "$ax = bx$ and the generalized eigenproblem," *Siam Journal on Numerical Analysis*, vol. 7, no. 4, pp. 479–492, 1970.
24. G. W. Stewart and J. G. Sun, "Matrix perturbation theory," 1990.
25. A. Krizhevsky, I. Sutskever, and G. E. Hinton, "Imagenet classification with deep convolutional neural networks," in *International Conference on Neural Information Processing Systems*, 2012, pp. 1097–1105.
26. C. Jiang, C. Yan, R. Yong, and K. J. R. Liu, "Maximizing network capacity with optimal source selection: A network science perspective," *IEEE Signal Processing Letters*, vol. 22, no. 7, pp. 938–942, 2014.
27. K. Arulkumaran, M. P. Deisenroth, M. Brundage, and A. A. Bharath, "A brief survey of deep reinforcement learning," 2017.
28. D. Drutskoy, E. Keller, J. Rexford, Scalable network virtualization in software-defined networks, IEEE Internet Computing 17 (2) (2013) 20–27.
29. N. Zhang and H. P. Yao, "Overview of euicc remote management technology," *Telecom Engineering Technics & Standardization*, 2012.
30. R. Jain, S. Paul, Network virtualization and software defined networking for cloud computing: a survey, Communications Magazine IEEE 51 (11) (2013) 24–31.
31. A. Fischer, J. F. Botero, M. T. Beck, H. D. Meer, X. Hesselbach, Virtual network embedding: A survey, IEEE Communications Surveys & Tutorials 15 (4) (2013) 1888–1906.

32. C. Liang, F. R. Yu, Wireless network virtualization: A survey, some research issues and challenges, IEEE Communications Surveys & Tutorials 17 (1) (2015) 358–380.
33. N. M. K. Chowdhury, R. Boutaba, Network virtualization: state of the art and research challenges, IEEE Communications magazine 47 (7) (2009) 20–26.
34. H. Zhang, C. Jiang, J. Cheng, and V. C. M. Leung, "Cooperative interference mitigation and handover management for heterogeneous cloud small cell networks," *Wireless Communications IEEE*, vol. 22, no. 3, pp. 92–99, 2015.
35. N. M. M. K. Chowdhury, R. Boutaba, A survey of network virtualization, Computer Networks 54 (5) (2010) 862–876.
36. Y. Zhu, M. Ammar, Algorithms for assigning substrate network resources to virtual network components, in: INFOCOM 2006. IEEE International Conference on Computer Communications. Proceedings, 2007, pp. 1–12.
37. M. Yu, Y. Yi, J. Rexford, M. Chiang, Rethinking virtual network embedding: substrate support for path splitting and migration, Acm Sigcomm Computer Communication Review 38 (2) (2008) 17–29.
38. L. Xu, C. Jiang, J. Wang, J. Yuan, and Y. Ren, "Information security in big data: Privacy and data mining," *IEEE Access*, vol. 2, no. 2, pp. 1149–1176, 2017.
39. N. M. M. K. Chowdhury, M. R. Rahman, R. Boutaba, Virtual network embedding with coordinated node and link mapping, Proceedings - IEEE INFOCOM 20 (1) (2009) 783–791.
40. C. Jiang, N. C. Beaulieu, Z. Lin, R. Yong, M. Peng, and H. H. Chen, "Cognitive radio networks with asynchronous spectrum sensing and access," *Network IEEE*, vol. 29, no. 3, pp. 88–95, 2015.
41. M. A. Ying-Jie, Z. Zhou, H. E. Wen-Cai, J. Zhang, and H. P. Yao, "Cognitive uwb orthogonal pulses design and its performance analysis," *Transactions of Beijing Institute of Technology*, vol. 31, no. 5, pp. 583–588, 2011.
42. S. Shanbhag, A. R. Kandoor, C. Wang, R. Mettu, T. Wolf, Vhub: Single-stage virtual network mapping through hub location, Computer Networks 77 (2015) 169–180.
43. D. Silver, A. Huang, C. J. Maddison, A. Guez, L. Sifre, G. Van Den Driessche, J. Schrittwieser, I. Antonoglou, V. Panneershelvam, M. Lanctot, et al., Mastering the game of go with deep neural networks and tree search, Nature 529 (7587) (2016) 484–489.
44. C. Jiang, C. Yan, K. J. R. Liu, and R. Yong, "Optimal pricing strategy for operators in cognitive femtocell networks," *Wireless Communications IEEE Transactions on*, vol. 13, no. 9, pp. 5288–5301, 2014.
45. V. Mnih, K. Kavukcuoglu, D. Silver, A. A. Rusu, J. Veness, M. G. Bellemare, A. Graves, M. Riedmiller, A. K. Fidjeland, G. Ostrovski, Human-level control through deep reinforcement learning, Nature 518 (7540) (2015) 529.
46. S. Mozer, M. C, M. Hasselmo, Reinforcement learning: An introduction, Machine Learning 8 (3-4) (1992) 225–227.
47. L. Meng, F. R. Yu, P. Si, H. Yao, E. Sun, and Y. Zhang, "Energy-efficient m2m communications with mobile edge computing in virtualized cellular networks," in *IEEE International Conference on Communications*, 2017.
48. C. Jiang, C. Yan, and K. J. R. Liu, "Evolutionary dynamics of information diffusion over social networks," *IEEE Transactions on Signal Processing*, vol. 62, no. 17, pp. 4573–4586, 2014.
49. X. Jin, P. Zhang, and H. Yao, "A communication framework between backbone satellites and ground stations," in *International Symposium on Communications & Information Technologies*, 2016.
50. S. Haeri, L. Trajkovic, Virtual network embedding via monte carlo tree search., IEEE Transactions on Cybernetics (99) (2017) 1–12.
51. R. Mijumbi, J. L. Gorricho, J. Serrat, M. Claeys, F. D. Turck, S. Latre, Design and evaluation of learning algorithms for dynamic resource management in virtual networks, in: Network Operations and Management Symposium, 2014, pp. 1–9.
52. H. Zhang, C. Jiang, N. C. Beaulieu, X. Chu, X. Wen, and M. Tao, "Resource allocation in spectrum-sharing ofdma femtocells with heterogeneous services," *IEEE Transactions on Communications*, vol. 62, no. 7, pp. 2366–2377, 2014.

53. S. Hougardy, The Floyd–Warshall algorithm on graphs with negative cycles, Information Processing Letters 110 (8) (2010) 279–281.
54. M. Thomas, E. W. Zegura, Generation and analysis of random graphs to model internetworks, College of Computing volume 63 (4) (1994) 413–442(30).
55. M. Abadi, A. Agarwal, P. Barham, E. Brevdo, Z. Chen, C. Citro, G. S. Corrado, A. Davis, J. Dean, M. Devin, et al., Tensorflow: Large-scale machine learning on heterogeneous distributed systems, arXiv preprint arXiv:1603.04467.
56. Q. Chao, C. Zhao, H. Yao, F. Xu, and F. R. Yu, "Why did you opt to switch off me? big data for green software defined networking," in *Globecom Workshops*, 2017.
57. L. Bottou, Online algorithms and stochastic approximations, in: D. Saad (Ed.), Online Learning and Neural Networks, Cambridge University Press, Cambridge, UK, 1998, revised, October 2012.

Chapter 6
Intention Based Networking Management

Compared to traditional networking which using command-line interfaces, intent-based networking abstracts network complexity and improves automation by eliminating manual configurations. It allows a user or administrator to send a simple request—using natural language—to plan, design and implement/operate the physical network which can improve network availability and agility. For example, an IT administrator can request improved voice quality for its voice-over-IP application, and the network can respond. For intent-based networking, the translation and validation system take a higher-level business policy (what) as input from end users and converts it to the necessary network configuration (how) by natural language understanding technology. In this chapter, we focus on how artificial intelligence technology can be used in the natural language understanding in translation and validation system. We firstly propose an effective model for the similarity metrics of English sentences. In the model, we first make use of word embedding and convolutional neural network (CNN) to produce a sentence vector and then leverage the information of the sentence vector pair to calculate the score of sentence similarity. Then, we propose the SM-CHI feature selection method based on the common method used in Chinese text classification. Besides, the improved CHI formula and synonym merging are used to select feature words so that the accuracy of classification can be improved and the feature dimension can be reduced. Finally, we present a novel approach which considers both the semantic and statistical information to improve the accuracy of text classification. The proposed approach computes semantic information based on HowNet and statistical information based on a kernel function with class-based weighting.

H. Yao et al., *Developing Networks using Artificial Intelligence*,
Wireless Networks, https://doi.org/10.1007/978-3-030-15028-0_6

6.1 CNN Based Sentence Similarity Model

Sentence similarity measurement is a fundamental problem in the research of natural language processing (NLP) and has been used in many language processing tasks such as query ranking, question answering, paraphrase identification and paraphrase generation, to cite a few [57]. Understanding the similarity between sentences is not a simple task for machines, because both semantic information and syntactic information should be taken into consideration. The distributed word embedding has been explicitly encoded many linguistic regularities and patterns in the works of Mikolov et al. [1–3] and others, and it has been widely utilized to extract semantic features of sentences. The syntax parsing is a challenge task due to its complexity, therefore it is a good idea that makes machines hidden learn grammar rules from corpus. Recently, as the increasing interest in neural networks, the idea is realized by means of neural network-based models.

In 2008, Collobert and Weston [5] proposed a convolutional neural network-based architecture for modeling sentences and used it in several learning tasks such as named entity tags, semantic roles, semantically similar words and so on. In their studies, words were embedded into the multi-dimension space. Inspired by their works, other researchers utilized word embedding and convolutional neural network in paraphrase identification [6, 7], sentences classification [4, 10, 12, 13] and semantic similarity measurement [8, 10, 14, 16]. Recurrent neural network, a time sequence neural network with variable-length input, is good choice for sequence modeling tasks like modeling sentences. Socher et al. [19] applied RNN in their task that finding and describing images with sentences. The disadvantage of RNN is its gradients vanishing over long sequence. To avoid the vanishing gradient problem, the long short-term memory networks (LSTM) were proposed, which is better in learning long range dependencies. Tai et al. [10] and Mueller et al. [20] used LSTM for sentence similarity measurement.

However, both long short-term memory networks and deep convolutional neural networks have a disadvantage of time-consuming when they are utilized for semantic similarity measurement [11]. In addition, deep convolutional neural networks need to be modified for measuring sentence similarity. The structures of them universally are complicated. Since the measurement of sentence similarity is a fundamental task in NLP and more commonly serves as a subtask that aids in solving other larger tasks, these models that have complicated structure and cost long time for training will consume extra resources that assigned for other subtasks and slow down the proceeding of the whole task [15].

In this chapter, we propose a novel model with word embedding and convolutional neural network for sentence similarity measurement. The major contributions of this part are as follows:

1. To reduce the training time and simplify architecture structure, we propose a novel sentence similarity model that is different from traditional long short-term memory networks and deep convolutional neural networks. The main body of our model is a shallow convolutional neural network without requiring

fully connected layer meanwhile modifying its structure targeting at the task of sentence similarity measurement.

2. The sentence with variable length is embedded into a multiple dimensional space. We extend the method of word embedding from word-level to sentence-level, with the aim of putting word similarity computation methods into sentence similarity measurement. Additionally, our model can handle any length sentences without needing clipping and padding operations.
3. We evaluate our model using different evaluation metrics with two different kinds of tasks, namely, semantic relatedness task (SemEval 2014, Task 1) and the Microsoft research paraphrase identification task. The obtained results achieve a good performance on both tasks.

6.1.1 The Proposed Model WSV-SCNN-SV

In this part, we propose our neural network-based model namely "the word set vectors-the shallow convolutional neural network-the sentence vector" (WSV-SCNN-SV). The main body of the model is the shallow convolutional neural network which takes the group of "word set vectors" as the input feature map and outputs the sentence representation—the sentence vector. The "word set vector" is a new idea considering as the improvement of the word embedding. Compared with word embedding, it contains more sematic information of sentences. We compute the similarity based on the sentence vectors that learned by the convolutional neural network instead of employing the convolutional neural network [17].

The organization of this subsection is summarized as follows: First, we introduce the whole architecture of our model, then we elaborate on each part of our model in detail.

6.1.1.1 The Overall Architecture of the Model

The hybrid architecture of WSV-SCNN-SV is shown in Fig. 6.1. In the architecture, there are four parts:

Part a: Sentences Pre-processing Since the raw sentence is not accepted by machines, it should be processed and encoded before it input into machines. The pre-process of the sentences includes word segmentation, removal of stop words, stemming and son on. In our work, we use word embedding to encode sentence in the way that words from each sentence are represented by their corresponding word vector after the pre-processing.

Part b: The Input Model We propose a novel method that transforms the sequence of word vectors into a three-dimensional tensor. The tensor which contains rich semantic and syntactic information of the sentence is the input feature map of the next convolutional neural network.

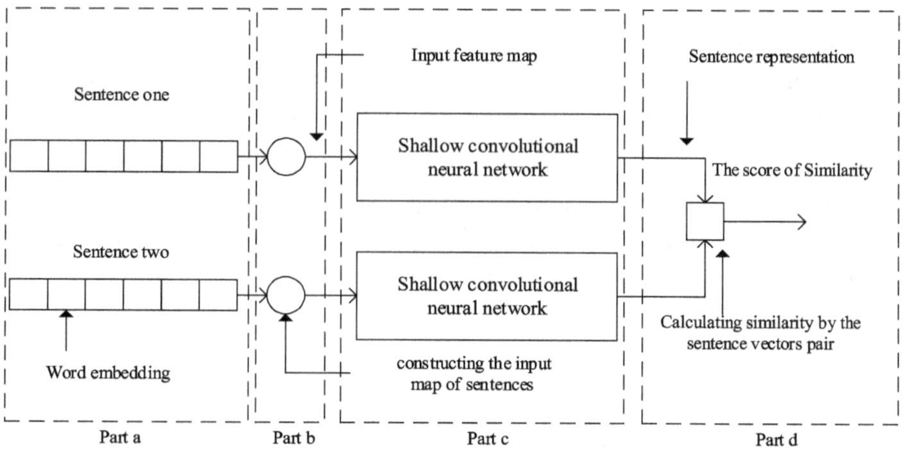

Fig. 6.1 The architecture of the model

Part c: Our Shallow Convolutional Neural Network The work of our convolutional neural network is learning the semantic and syntactic features from the input tensor and producing the sentence representation. We make some modification to the convolutional neural network. First, we remove fully connected layer from our model. After flowing over several convolutional and pooling layers, the tensor is converted into a vector which extracts rich features from the input tensor. Second, we apply both k-max pooling and max pooling operation in the model. Third, we make our convolutional neural network support the input feature map with unfixed size by means of k-max pooling operation.

Part d: Similarity Computation The sentence vector pair is used to compute the score of similarity. Many methods are available for similarity measurement between two vectors. In our work, we use cosine distance, Euclidean distance and Manhattan distance, respectively.

In summary, the aim of the first three parts of the model is to accomplish the task that embeds sentences into a high dimensional space. The goal of the last part of our model is to calculate the score of similarity with the sentence vector pair.

6.1.1.2 Sentences Pre-processing

Since each word is an atomic semantic unit for a sentence, it is of great significance for machines to capture fine-grained features from sentences [61]. Different sentences may use different words to convey the high similar information, for instance, "He studies computer science in college" and "His major is CS". Many remarkable researches have been done in modeling words like distributed word embedding which explicitly encode many linguistic regularities and word similarities.

One of notable works on distributed word embedding is word2vec model [1, 3]. In word2vec model, the result of a vector calculation vector("Madrid") − vector("Spain") + vector("France") is closer to vector("Paris") than to any other word vector. It is universal in related papers that base on word2vec to model sentence measuring similarity. Similarly, in this part, we apply pre-trained word embedding to model words.

There are many simple methods to produce the sentence representation using word embedding. Assume $\mathbf{w}_i = (w_i^1, w_i^2, \ldots, w_i^d)$ is a d-dimensional word vector corresponding to the i-th word in the sentence. A sentence contained n words can be represented as follows:

$$S = \frac{1}{n}\sum_{i=1}^{n} \mathbf{w}_i \tag{6.1}$$

$$S = (\mathbf{w}_1^T, \mathbf{w}_2^T, \ldots, \mathbf{w}_n^T) \tag{6.2}$$

In prior studies, these sentence representations are input into neural network to finish final task. Formula (6.1) uses the average value of these word vectors to express the sentence [23, 25–27, 30, 31], despite this process is very simple, its performance in practice is not poor. Formula (4.31) concatenates word vectors to produce a two-dimensional matrix to express the sentence, which is a common method to represent sentence. Different from formula (4.30), the method of formula (4.31) keeps order information of words. Kim [13] used the method, and Hu et al. [6] made some improvement of formula (6.2).

However, our idea is different from the previous work. Our motivation is that sentence representation should learn from neural network as the word embedding does. After words in the sentence are represented by their corresponding word vector, we construct a three dimensional tensor and then put the tensor into the convolutional neural network to learn the sentence representation.

6.1.1.3 The Input Model Based on "Binary-Gram Word Set Vector"

The word is not the only factor to be taken into consideration in the sentence representation. The dependencies between word and phrase, phrase and phrase cannot be ignored. The representation of the sentence in formula (6.2) can be treated as a feature map with one width, n height and d channels. However, the disadvantage of the representation is that long-range semantic dependencies between words have been ignored.

To consider the long-range semantic dependencies in our model, we propose a new vector dubbed "word set vector". The idea of word set vector is inspired by the idea that sentences can be represented by the average of word vectors. We extend this idea from sentences to phrases and collections of words that are semantic dependencies with each other. Word set vector $\overline{\mathbf{w}}$ composed with j word vectors

$\mathbf{w}_{l_1}, \mathbf{w}_{l_2}, \ldots, \mathbf{w}_{l_j}$ is defined as follows:

$$\overline{\mathbf{w}} = \sum_{i=1}^{j} \lambda_i \mathbf{w}_{l_i} \tag{6.3}$$

where the sequence $\{l_1, l_2, \ldots, l_j\}$ denotes a subsequence of the sequence $\{1, 2, \ldots, n\}$, λ denotes the weight of word vector ($\sum_{i=1}^{j} \lambda_i = 1, \lambda_i > 0$).

The word set vector composed by word vectors $\mathbf{w}_{l_1}, \mathbf{w}_{l_2}, \ldots, \mathbf{w}_{l_j}$ is order sensitive. The word set vector $\overline{\mathbf{w}}$ composed with j word vectors (allowed repeated word vectors) is defined as j-gram word set vector. We deem that the parameter λ should be trained by machines, it can be calculated as follows:

$$\lambda_i = \frac{exp(\alpha_i)}{\sum_j exp(\alpha_j)}, \alpha_j \in \mathbb{R} \tag{6.4}$$

The word set vector uses the weight λ to punish disorder word vectors. A sentence with n words has n^j j-gram word set vectors. A sentence with n words has n^2 d-dimensional binary-gram word sets vectors which can be transformed into a feature map with n width, n height and d channels. The binary-gram word set vector is calculated by:

$$\widetilde{\mathbf{w}}_{l_1 l_2} = \lambda_1 \mathbf{w}_{l_1} + \lambda_2 \mathbf{w}_{l_2} \tag{6.5}$$

where $l_1, l_2 \in \{1, 2, \ldots, n\}$, $\lambda_1 + \lambda_2 = 1, \lambda_1, \lambda_2 > 0$. If $l_1 = l_2$, $\widetilde{\mathbf{w}}_{l_1 l_2}$ denotes a binary-gram word set vector. The weight λ_1, λ_2 are calculated by gradient descent.

Algorithm 1 shows how the input feature map is composed. The units of the map are these binary-gram word set vectors in the sentence.

Algorithm 1 The computation process of the input feature map

Input: Word embeddings, $w_1, w_2, \ldots, w_n$
Output: The input feature map, map
Initialize the input feature map: $map \in \mathbb{R}^{n \times n \times d}$ **to all zero**
Initialize variables: α_1, α_2 **to 0.4,0.6**
$\lambda_1 = exp(\alpha_1)/(exp(\alpha_1) + exp(\alpha_2))$
$\lambda_2 = exp(\alpha_2)/(exp(\alpha_1) + exp(\alpha_2))$
for $i = 1$ to n **do**
    for $j = 1$ to n **do**
        for $depth = 1$ to d **do**
            $map[i][j][depth] = \lambda_1 \cdot w_i[depth] + \lambda_2 \cdot w_j[depth]$
        end for
    end for
end for

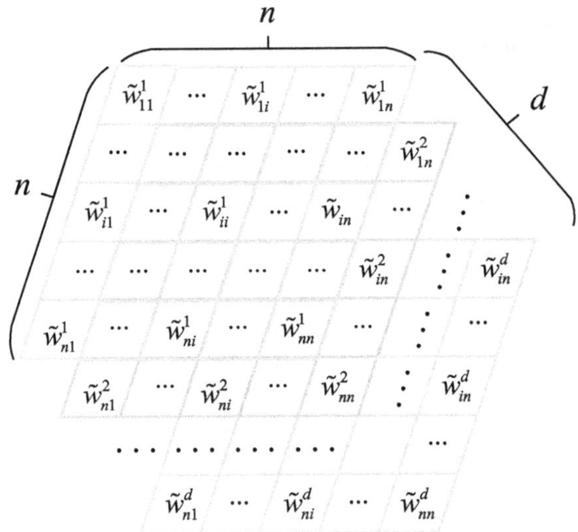

Fig. 6.2 The input feature map of the convolutional neural network where $\tilde{w}_{ij}^k = \lambda_1 w_i^k + \lambda_2 w_j^k$ denotes the k-th dimensional value of word set vector that is composed of word vector $\mathbf{w}_i$ and $\mathbf{w}_j$

Figure 6.2 illustrates the feature map that is composed of n^2 word set vectors. It is one of creative works in the part that extending the feature map of sentences from a two-dimensional matrix to a three-dimensional tensor. As the feature map is the same as the image data in form, we hope that the next convolutional neural network can fully extract sentence features from the input feature map as it does on images.

6.1.1.4 Our Shallow Convolutional Neural Network

The complete multi-layer's convolutional neural network should contain fully connected layer just like Fig. 6.3a. In our convolutional neural network, we remove the fully connected layer from the network as it is not essential for learning sentence vectors and consumes time for training. Our proposed convolutional neural network only contains layers that consist of convolutional sublayers and pooling sublayers. In addition, a sum function that computes the sentence vectors is applied at the output of the last layer.

Convolutional Sublayers The formula of convolution is as follows:

$$X_j^l = \phi(\sum_{i \in M_j} X_i^{l-1} * K_{ij}^l + b_j^l) \tag{6.6}$$

where X_j^l denotes the j-th map on the l-th sublayer, $\phi(x)$ denotes the active function, M_j denotes the number of maps on the $(l-1)$-th sublayer, K^l denotes the filter of the l-th sublayer, b_j^l denotes the bias.

We do not make any modification to convolutional sublayers. As can be seen in Fig. 6.3a, zero padding (dashed blocks are zero padding) is used in the convolutional

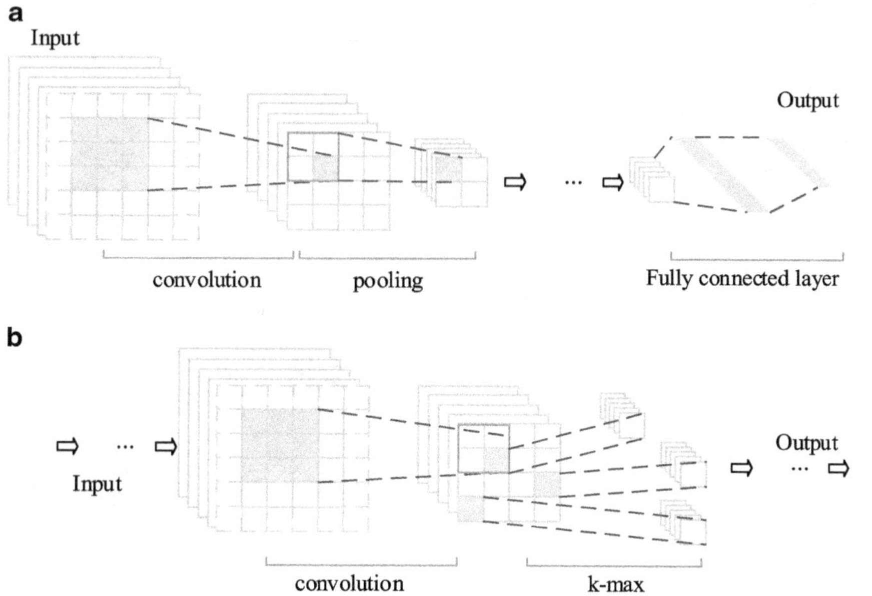

Fig. 6.3 The convolutional neural network, k-max pooling. (**a**) The convolutional neural network with max-pooling operation. (**b**) The convolutional neural network with k-max pooling operation

operation so that the size of input feature map of convolutional layer is the same as the size of output feature map of the convolutional layer.

The activate function of convolutional sublayers is ReLU function:

$$\phi(x) = \begin{cases} x \text{ if } x \geq 0 \\ 0 \text{ if } x < 0 \end{cases} \tag{6.7}$$

Though, the performance of ReLU function is poor than sigmoid function's, it is simple and takes less training time to achieve convergence.

Max Pooling Sublayers Pooling sublayers are followed convolutional sublayers. The pooling operation is also called down sampling. In pooling operation, both width dimension and height dimension of the feature map will be reduced. However, the depth dimension of the feature map remains unchanged. Taking 2 × 2 pooling window for example (it can be seen in Fig. 6.3a), the width and height of feature map will reduce half after a pooling operation. If the size of input feature map of multi-layer network is defined, after several times pooling operation, the size of output feature map is also definite.

The formula of convolution is as follows:

$$X^l_j = \beta^l_j down(X^{l-1}_j) + bias^l_j \tag{6.8}$$

where β_j^l denotes the weight, $down(x)$ denotes the down sampling, $bias_j^l$ denotes the bias.

The k-Max Pooling Sublayer The length of the sentence is variable. To support sentences with different length, the size of input layer of the convolution neural network must be changeable. To remain the same size of output layer of the convolutional neural network with variable input feature map, k-max pooling is used on the last pooling layer to replace max pooling operation. Figure 6.3b shows k-max pooling operation ($k = 3$) on the input feature map. The pseudo code of k-max pooling is as follows:

Algorithm 2 The computation process of the k-max pooling

```
Input: the three-dimensional feature map, tensor ∈ ℝ^(width×height×depth)
Output: The k vectors, vectors ∈ ℝ^(k×depth);
function kMax(tensor)
    tensor = reshape(tensor, [width × height, depth])
    //After the reshape operation, tensor is a two-dimensional matrix: tensor ∈ ℝ^((width×height)×depth)
    for i = 1 to tensor.depth do
        tensor[:][i] = AscendingSort(tensor[:][i])
    end for
    for i = 1 to k do
        vectors[i][:] = tensor[i][:]
    end for
    return vectors
end function
```

The k-max pooling used in our model is different from the k-max pooling used in Nal et al.[12]. In Nal et al.[12], after k-max pooling, the relative position of data in the original feature map is preserved. In contrast, we discard the relative position of data because we subsequently summed the data up to calculate the sentence vector. In Fig. 6.3b, the k-max operation on any $n \times n \times d$ ($n \geq 2$) input feature map produces three d-dimensional vectors.

The Sum Function Taking the sentence "The cat sits on the mat" for example, its subject, predicate and object convey enough meaning to imply the whole sentence. Similarly, we hope that words or phrases that convey the meaning of sentence can be "extracted" from the sentence by means of k-max pooling operation. The k vectors outputted by k-max pooling is called the vector group of the sentence. There is no fully connected layer in our network. The sentence vector is the average value of the vector group of the sentence and is calculated by:

$$v_S = \frac{1}{k}\sum_{i=1}^{k} u_i \tag{6.9}$$

where v_S represents the sentence vector, $u_1, \ldots, u_i, \ldots, u_k$ denotes the vector group of the sentence.

Algorithm 3 shows the pseudo code of the shallow convolutional neural network.

Algorithm 3 The computation process of the shallow convolutional neural network

Input: the input feature map, map_0
Output: The sentence vector, v_s
Initialize the filter: $filter^1, filter^2, \ldots, filter^l$
for $i = 1$ to $l - 1$ **do**
  $conv_i = ReLu(convolute(map_{i-1}, filter^i, padding =' same'))$
  $map_i = MaxPooling(conv_i)$
end for
$conv_l = ReLu(convolute(map_{l-1}, filter^l, padding =' same'))$
$vectors = kMax(conv_l)$
$v_s = average(vectors[1], vectors[2], \ldots, vectors[k])$

In our application, the number of layers in our network is no more than three. The work of fully connected layers is substituted by the next calculation unit. Simple in structure and fast in training are the advantages of our model. The model with two layers of network is shown in Fig. 6.4.

6.1.1.5 Similarity Computation

The similarity of sentences is calculated by the sentence vector pair [63]. There are many methods to compute the similarity of vectors. We choose cosine distance, Euclidean distance and Manhattan distance to evaluate the score of similarity. However, the value range of Euclidean distance and Manhattan distance is not [0,1]. We need to make modification to them. The score calculated by Euclidean distance and Manhattan distance is as follows:

$$score = f(||v_{s1} - v_{s2}||), score \in [0, 1] \tag{6.10}$$

where f denotes a monotone decreasing function. And the codomain of f is [0, 1]. We use three different f in the model: $f_1(x) = e^{-x}$, $f_2(x) = \frac{1}{1+e^{-x}}$ and $f_3(x) = \frac{1}{1+x}$. Since the output of ReLu function are not negative, there is no need to concern the negative value of cosine distance. The cosine distance of the sentences vector pair can be directly used as the score of similarity.

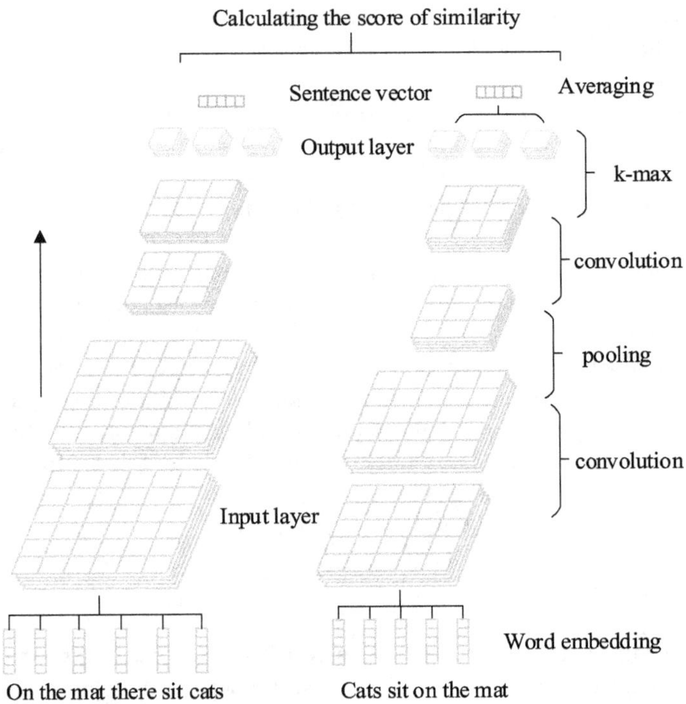

Fig. 6.4 The sentence similarity model

6.1.2 Experimental Results and Analysis

In the subsection, we present detailed experimental setup and loss functions in our training process, after which experimental analysis is illustrated respectively. Two different data set SICK and MSRP are used to evaluate our model.

6.1.2.1 Experimental Setup

We did all the experiments on a personal computer with 8 gigabytes memory and Intel i7 quad core CPU. In experiments, we use GloVe word embedding [66] (trained on Wikipedia 2014 + Gigaword 5). And we fine-tune the word embeddings on training. Words which not present in the GloVe are initialized randomly.

If padding is not allowed in pooling sublayer and k-max pooling sublayer, the length of sentence must meet a lower limit. The length of the sentence satisfies:

$$Lengths_S \geq (f_w)^{L-1}k \tag{6.11}$$

where f_w represents the width of pooling window (In this part, the width of pooling window is equal to its height), k denotes the parameter of k-max, L denotes the number of hidden layers.

In the model, the number of channels of input layer is the same as the dimension of word embedding (commonly more than 50-dimension). Adding a new hidden layer to network will greater increase the number of training parameters in the model. To limit the number of parameters, the number of layers L is no more than three.

On training task, to keep the same size of input map in the same batch, we have to use zero padding in input layer. However, on testing task, the model supports the different size of input map.

The model contains many hyper parameters. This is a shortcoming of the model. There are the number of layers L, the parameter k, the width of convolutional filter, the width of pooling window and the number of hidden feature maps to be artificially specified. But the model with non-fully connected layer and fewer hidden layers is less than most of deep neural networks on the total amount of parameters.

6.1.2.2 Training

The data set of our experiments are SemEval-2014 Sentences Involving Compositional Knowledge (SICK) data and Microsoft research paraphrase identification (MSRP)[32, 35]. SemEval (Semantic Evaluation) is a computational semantic analysis system organized by the Special Interest Group on the Lexicon of the Association for Computational Linguistics. SICK is the data set of SemEval-2014 [36] Task 1 competition, the aim of which is evaluated compositional distributional semantic models on full sentences through semantic relatedness and entailment. Many participants submitted their works on the task and their results are available in Marelli et al. [37]. SICK contains training data (4500 sentence pairs), trial data (500 sentence pairs) and test data (4927 sentence pairs). Relatedness score of sentence pair can range from 1 (completely unrelated) to 5 (very related). MSRP contains training data (4076 sentence pairs) and test data (1725 sentence pairs). Relatedness score of sentence pair is one or zero.

Hyper parameters of the model were set as follows: parameter k of k-max pooling was 3; the size of convolutional filter was 3×3; the single training batch size was set to be 50. We are fine tuning the word embedding on both tasks. The loss function of these two tasks are different.

On the SICK data set, we train model to minimize the loss function of mean squared errors (MSE):

$$loss_1 = \frac{1}{m}\sum_{i=1}^{m}(sim_p - sim_l)^2 \tag{6.12}$$

where sim_p denotes the predicted value of similarity score, sim_l represents similarity score that is calculated as the average of ten human ratings collected for each pair, m is the scale of training data set.

Besides, to explore the influence of loss function on the result, we use the other loss function—KL divergence loss:

$$loss_2 = \frac{1}{m}\sum_{i=1}^{m}[q \cdot log\frac{q}{p} + (1-q) \cdot log\frac{(1-q)}{(1-p)}] \tag{6.13}$$

where p is normalized sim_p and q is normalized sim_l. Considering denominator p and $1 - p$ could be zero, Laplace Smoothing is being used on this loss function.

On the MSRP data set, we training model to minimize cross entropy between the predicted value and the label.

6.1.2.3 Experiments Analysis

Figure 6.5 illustrates results on the SICK test set. The six results at the bottom of the figure are the best six results released by SemEval. These results represent the high level of traditional methods. It can be seen that the result of this model is ranked third in above results, according to Pearson coefficient (SemEval official ranking). Similarly, according to Spearman coefficient, the result is ranked the third. And the

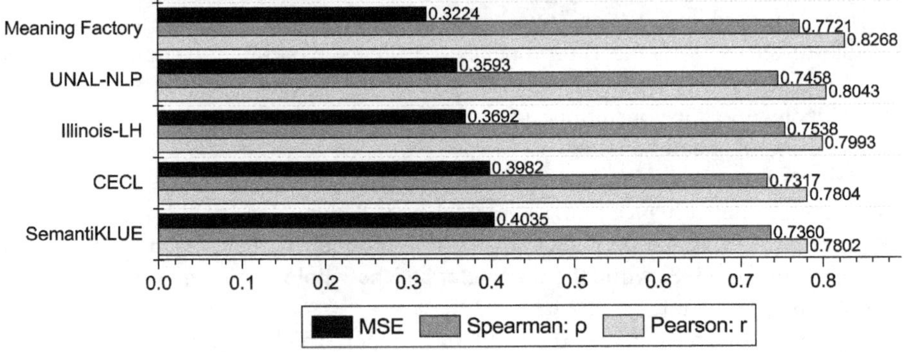

Fig. 6.5 The test result on SICK data set. DT-RNN [19], SDT-RNN [19], ECNU [38], Meaning Factory [40], UNAL-NLP [41], Illinois-LH [42], CECL [43], SemantiKLUE [45]

result on MSE is ranked the fourth, very closed to the third 0.3593. There is no distinct difference between the result of our model and the best results. Differences between our result with the best on Pearson, spearman and MSE respectively are 0.0159, 0.0137 and 0.0382. Figure 6.5 also provides a comparison of the result of this work with results of two different RNN models. Compared to RNN, it is disadvantaged for the CNN network to deal with long-term and short-term dependencies between words in the sentence. Surprisingly, the test result of the model is slightly better than results of these two RNN models. But the results of our model are not as good as the results of literatures (He et al. [14], He and Lin [16]). The results of their work on Pearson coefficient are 0.8686 and 0.8784 respectively. Despite the performance of our model is poorer than these deep neural networks, the work of us is still valuable: Compared with traditional methods, our work achieves better performance. And compared with deep neural networks, our model is very practical in applications since it is simple in architecture and fast in the training.

Table 6.1 presents the methods and resources used by several models that well-performed on the SICK data set. It can be seen that the well-performed ECNU model uses four learning methods, WordNet and additional Corpus; the second ranked model, The Meaning Factory, also uses three different resources including WordNet. In contrast, our model only uses word embeddings and convolution neural networks. Our model is much simpler in structure, but can achieve the same degree of test performance.

Figure 6.6 demonstrates the training curve of our model on SICK set (Learning rate is 10^{-3}). From Fig. 6.6, it can be seen that the performance of these two loss functions has no significant difference on the test results. The results show that both curves are fast convergence.

We used different dimensional word embeddings to test our model, and recorded their training time. We chose a single-layer network to accomplish the experiments. Table 6.2 shows the corresponding results. All of experiments that list the training time are done on the same computer.

As can be seen from Table 6.2, the result of a single-layer network which contains 1.62 million parameters and spent around 80 min on training can rank the third in Fig. 6.5. After reducing the half number of parameters, the test result does not show obvious difference with the same epoches. The Pearson correlation coefficient of the result reaches 0.793, the fourth level in Fig. 6.5, and are very closed to that of the third. Meanwhile, the consuming time can be reduced to a half: the training process can be accomplished in 37 min.

Though the model of the third line in the table (100-dimensional word vector) has not bright performance under ten epoches, if the epoches are 40, the metrics of its Pearson correlation coefficient was 0.7868 around, its mean squared error was approximately 0.3878, while the parameters of the whole model are only 3/8 of the second case in the table.

Figure 6.7 shows that the result of our model which is pretrained on SICK corpus outperform others on MSRP test set. Meanwhile, the result of our model without pretraining does not perform well. Thus, pretraining is helpful to improve the performance of our model. We assume that our model may perform better after

Table 6.1 The methods and resources that models applied [37]

Model	Learning methods						External resources					
	SVM and kernel methods	K-nearest neighbours	Classifier combination	Random forest	Convolutional neural network	Other	WordNet	Paraphrases DB	Other corpora	ImageFlicker	STS MSR-video description	Word embeddings
ECNU	Y	Y	Y	Y			Y		Y			
The Meaning Factory				Y			Y	Y	Y			
UNAL-NLP						Y	Y	Y	Y			
Illinois-LH						Y	Y			Y	Y	
This work					Y							Y

Y represents the model used this method/resource

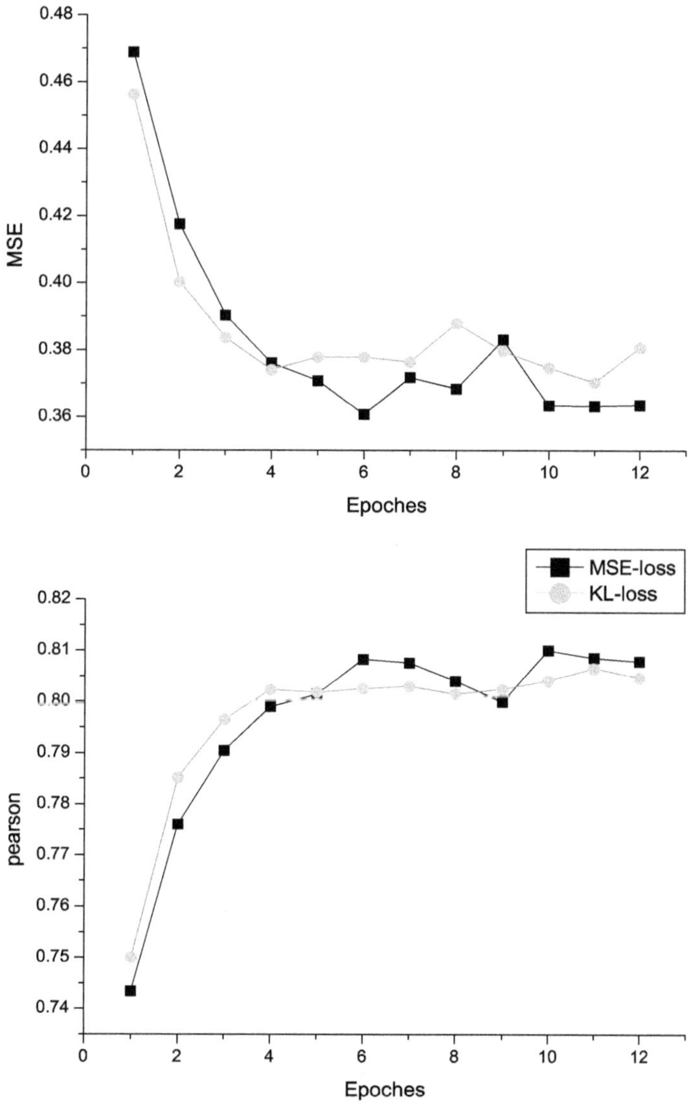

Fig. 6.6 The relatedness curve of mean square errors, Pearson coefficient and epoches on test data set using KL divergence loss function and MSE loss function

pretraining on a large amounts of appropriate training data, which we leave to our future work. (In Fig. 6.7, the work of Hu et al.[6] is based on CNN, others' work in Fig. 6.7 are based on traditional methods. The results of Baseline and Rus et al.[46] is reported by Hu et al.[6], and the result of Rus et al.[46] is the best one in the report.)

Table 6.2 The results on the different settings of the model

params	r	ρ	MSE	T
1,620,002 (300)	0.8069	0.7433	0.3897	4731
720,002 (200)	0.7930	0.7308	0.4269	2197
270,002 (100)	0.7679	0.7052	0.4419	819
90,002 (50)	0.7623	0.6999	0.5076	495

The "params" represents the parameters of the whole model to be trained, the content in the parentheses denotes the dimension of word vector. r denotes Pearson correlation coefficient, ρ denotes spearman correlation coefficient, MSE denotes mean square error, and T denotes the average time spent for the training (Unit: seconds). The epoches of the results are ten

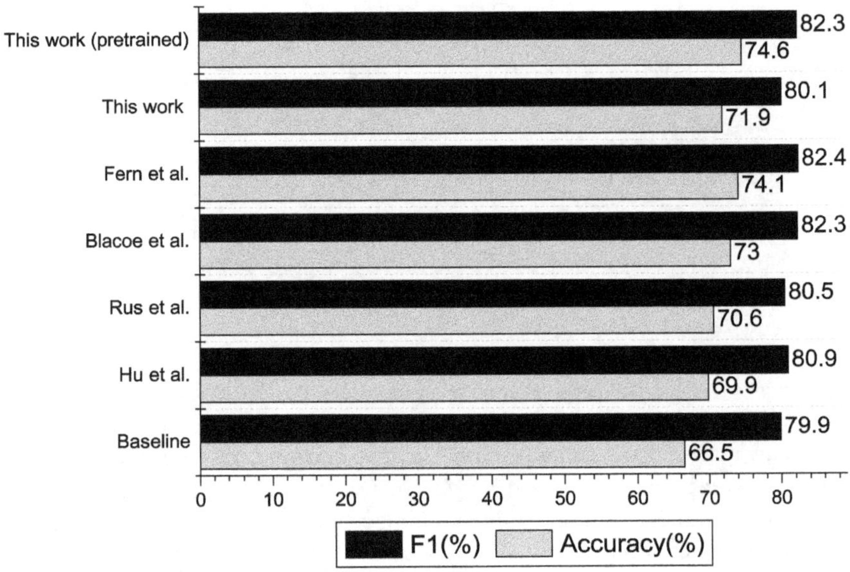

Fig. 6.7 The test result on MSRP data set Baseline[6], Hu et al.[6], Rus et al.[46], Blacoe et al.[47], Fernando and Stevenson [49]

6.2 A Feature Selection Method Based Text Classification System

With the development of the Internet, the amount of Chinese text information shows an exponential growth trend [50]. How to effectively manage the massive Chinese documents and mine the information contained in the documents has become a critical research problem. Automatic Text Classification can complete the work of text processing effectively. It also plays an important role in natural language processing (NLP) and data mining.

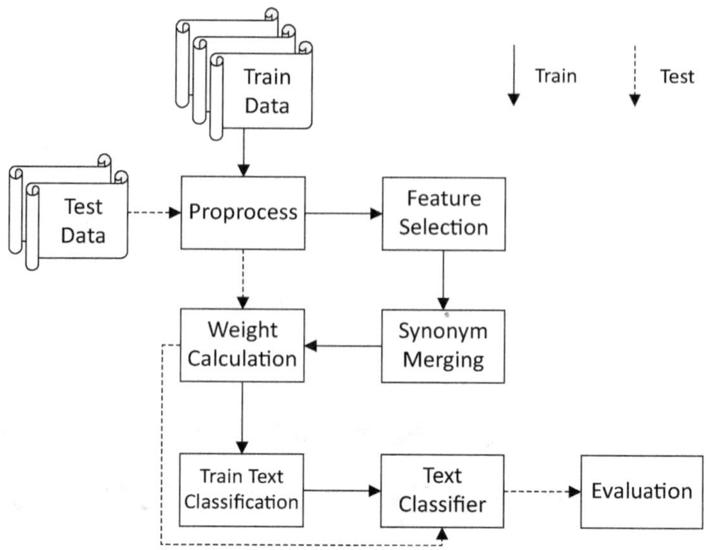

Fig. 6.8 Text classification model framework

The most common method used in text classification is the vector space model (VSM). It represents text as a feature vector. The specific process is shown in Fig. 6.8.

From the above figure, we know that the first step in Chinese text classification is to preprocess the text, including word segmentation, part of speech tagging, removal of stop words, etc. The purpose is to remove the useless words, and only leave the nouns, adjectives and verbs that contain category information. After this, the text can be represented as a vector to form VSM. Then we use the feature selection method to select the feature words that can symbolize the text categories, and merge the synonym to reduce dimensions. Next, TF-IDF [9, 18] method is used to calculate the weight of each feature of each text to transform the text into a feature vector. Last but not least, by using the Bayesian classifier to train the sample data, we can get the final text classifier.

Feature selection is the most important step because the selected feature words directly affect the accuracy of the classifier. In VSM, the best feature selection method is χ^2 statistics (CHI) [21, 52, 53]. But the defect is high-dimensional feature vectors selected by CHI may cause dimension disaster. The writer consider to merge the synonyms among the feature words selected by CHI so that the dimension of feature space can be reduced. Then in next step an improved TF-IDF method is used to calculate the feature weights for each word to generate the feature vector of each text. This part mainly studies the influence of feature selection and synonym merging on the accuracy of classification in automatic text classification. Synonym merging can reduce the dimension of the feature space and improve the classification performance. The study of feature selection algorithm and synonym merging has a strong practical significance. The main contributions of this part are as follows:

1. We presented a new feature selection algorithm named SM-CHI based on an improved CHI [22, 54] formula and synonym merging to achieve efficient feature selection and dimension reduction;
2. We found that the original CHI formula multiplied by a log term has the best classification performance comparing with the original CHI and two improved CHI algorithms [28, 54, 55];
3. The choice of thresholds α $(0 \leqslant \alpha \leqslant 1)$ is critical. Only the most similar feature words will be merged when α is close to 1, so we use grid search method to find the optimal α. The result show that the classification accuracy is highest when α is equal to 0.8;
4. We proposed three improved weight calculation methods based on TF-IDF. The experimental results show that using the maximum value of the synonym group as the feature weight is the best way.

6.2.1 *Text Classification Model Based on Semantic Similarity*

In this subsection, we mainly introduce the text classification model based on semantic similarity. Wherein, Sect. 6.2.1.1 describes the feature selection method based on χ^2 statistic. Section 6.2.1.2 introduces the method of synonym merging; Sect. 6.2.1.3 presents the traditional weight calculation method, TF-IDF.

6.2.1.1 Improved Feature Selection Algorithm

In text classification, a feature word and its category tend to obey the CHI formula. Higher CHI value implies that a feature word has stronger ability to identify a category. The CHI value of word is calculated as follows [53]:

$$\chi^2(t, c) = \frac{N * (AD - BC)^2}{(A + C)(A + B)(B + D)(C + D)} \tag{6.14}$$

where N is the size of the training set, A is the number of documents that belong to class c and contain the word t; B is the number of documents that do not belong to class c but contain the word t; C is the number of documents that belong to the class c but do not contain the word t; D is the number of documents that do not belong to class c and don't contain the word t.

Although the CHI formula has a relatively good performance in text classification, it also has some shortcomings [24, 59]. First of all, high-frequency words that appear in all categories have higher CHI values, but they do not make much sense for class distinctions. Secondly, the CHI formula only considers the appearance of a word but not the frequency of the word in a document. Therefore, CHI formula also has "low frequency words flawed". For example, assuming word $t1$ appears in 99 documents, each appears ten times; word $t2$ appears in 100 documents, each appears one times; obviously $t2$ has a higher CHI value, but in fact $t1$ is more representative for this category. There are many studies amended for its defects. The work in [54] proposed the multiplication by a log entry based on the original CHI to reduce the CHI value of high-frequency words. The formula is as follows:

$$chi\_imp\_1 = \log(\frac{N}{A+B}) * \chi^2(t, c) \tag{6.15}$$

where $A + B$ represents the number of documents that contain word t, and N represents the total number of documents. In this case, the CHI value of the high-frequency words that appear in all categories are close to zero so that they won't be selected as a feature word.

In addition, the work in [55] has made a corresponding improvement to the word frequency, which is multiplied by term $\beta(t, c)$ on the basis of the original CHI formula. The calculation formula is as follows:

$$chi\_imp\_2 = \beta(t, c)\chi^2(t, c) \tag{6.16}$$

where $\beta(t, c)$ is calculated as follows:

$$\beta(t, c) = \frac{tf(t, c)}{\sum_{i=1}^{m} f(t, c_i)} \tag{6.17}$$

In the formula, m is the total number of categories, $tf(t, c)$ is the frequency of the word t in the category c.

6.2.1.2 Synonym Merging Algorithm Based on "Tong YiCi Cilin"

Some of the feature words selected by CHI formula may be the same or have similar meaning. They have the same effect on class distinctions. If the synonym are merged, not only the classification accuracy will be improved, but also the dimension of the feature space can be reduced so that the efficiency of the algorithm can be improved. For example, "GanMao", "ZhaoLiang", "ShangFeng" are the synonym of "Cold" in Chinese. If the "Health Care" category articles contain these words respectively, then the feature words for the text classification contain too much redundant information. We use the method of synonym merging to deal with it. In this part, we use the "Tong YiCi Cilin" provided by Harbin Institute of Technology as the method of word similarity calculation [60]. Its structure has five layers. You can easily calculate the similarity between the two terms. The structure of "Tong YiCi Cilin" is shown in Fig. 6.9.

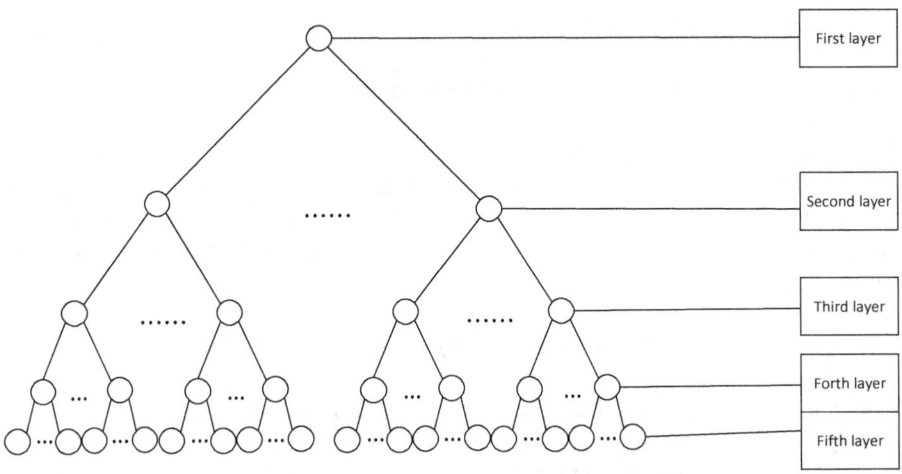

Fig. 6.9 Hierarchy structure of "Tong YiCi Cilin"

The concrete similarity calculation method is introduced in detail in [60]. When the similarity of two words is greater than threshold α, they will be regarded as a pair of synonym to merge. The optimal value of α will be discussed later in the experiment. In addition, all merged synonyms are stored in a list. Nested lists are used to store feature words so that all synonym information remains in the feature vector. To calculate the feature vector of each document, we propose three improved methods which will be discussed in Sect. 6.2.1.3.

6.2.1.3 Weight Calculation Method

Traditional TF-IDF weight calculation formula [9] is as follows:

$$weight_{t,d} = tf_{t,d} + idf_t \tag{6.18}$$

This formula represents the weight calculation method for word t in document d. Here, $tf_{t,d}$ denotes the frequency of occurrence of word t in d, and idf_t denotes the anti-document frequency of t, which is used to quantify the distribution of t in the training set. If n is used to denote the number of documents which contain t in the training set, the calculation formula of idf_t is as follows:

$$idf_t = \log(\frac{N}{n}) \tag{6.19}$$

As we mentioned above, the TF-IDF method is used to calculate the weight of each feature word in each text.

6.2.2 *Algorithm Description*

6.2.2.1 Feature Selection Method Based on the Synonym Merging

In this subsection, we introduce the feature selection algorithm (SM-CHI). This method firstly selects candidate feature words based on an improved CHI formula, and then merge synonyms to re-select those feature words that can represent the categories better and reduce dimension. The method is represented as the following formula:

$$SM - CHI = LF(t) * CHI(t) * SM(t) \tag{6.20}$$

where $LF(t)$ denotes whether the word t exists in the word bag or not, and is mainly decided according to the part of speech and stopping words. If the word t is a stopping word and in the part of speech that does not belong to verb, noun, adjective, $LF(t) = 0$, otherwise $LF(t) = 1$. $CHI(t)$ represents the CHI value of the word t, and is calculated by Eq. (6.15). $SM(t)$ indicates whether the word t contains synonym. If yes, it needs to merge all of its synonyms.

Firstly, all the texts in the training set are preprocessed, including Chinese word segmentation, part-of-speech tagging, and discarding stop words. The remaining words constitute the word bag of the training set. Secondly, we calculate the CHI value of each word. Choose the first 200 words from each category to form candidate sets of feature words. Note that the characteristic words selected for each category may be duplicated. The candidate set is stored using the HashSet (a data structure) and the de-emphasis is performed. After obtaining the candidate set, the similarity between each word is calculated according to "Tong YiCi Cilin" and threshold is set to α. The synonym merging is performed only when the word similarity is greater than α. We will experimentally determine the optimal value of hyper-parameter α. The pseudocode of SM-CHI is shown in Algorithm 4.

6.2.2.2 Improved Method for Calculating Eigenvalue Weight

In the scene of SM-CHI feature selection method presented in this part, the traditional TF-IDF formula has some drawbacks. For the features after synonym merging, the original weight calculation formula will cause "unfairness". Because the merged feature words are stored in the nested list, so which word among them will be regarded as the feature is a question. For this problem, we present the following three solutions:

- Sum the weights of all items up in the feature list of each dimension as the weight of the list;
- Take the largest weight among the synonym as the weight value of the feature;
- Multiply the first item by 1.1 for times of the number of items in the feature list.

Algorithm 4 SM-CHI feature selection algorithm

```
Input:a training set D
Output:a feature space F
(Initialization) A,B,TFIDF can all be a null dict
for each file in D:
    word_list=file.process()
    for word in word_list:
        A[file.class][word] += 1
        TFIDF[file.class][file.num][word] +=1
end for
Calculate B in the CHI formula from A
for cla in
    for word in cla:
        Calculate the CHI value according to formula (6.15)
    Selects the first 200 words as the feature
end for
Combine features of the 9 categories to obtain word_features
for word1,word2 in word_features:
    sim=calcWordsSimilarity(word1,word2)
    if sim > α:
        Merge word1 and word2
end for
```

6.2.3 *Experiments and Results*

In this subsection, the following three groups of experiments are carried out to test the three innovation points of SM-CHI with control variable method. In Experiment I, we test the three feature selection algorithms without using synonym merges. In Experiment II, we use the first improved CHI method to select features and use grid search method to find the optimal threshold α. On the basis of Experiment I and II, we designed Experiment III to find the best weight update method.

6.2.3.1 Performance Evaluation and Data Set

The standard precision rate P, recall rate R and $F1$ score are used to measure the classification performance. For the i_th category, the formula is as follows [62]:

$$P_i = \frac{TP}{TP + FN}, R_i = \frac{TP}{TP + FP}, F1_i = \frac{2 * P_i * R_i}{P_i + R_i} \tag{6.21}$$

where TP is the number of documents correctly classified as class i, FP is the number of documents classified as class i but not actually i, and FN is the number of documents that is not classified as class i but is actually class i.

This article will use the whole network news data set provided by Sogou Lab to test our experiments. The corpus includes nine kinds of news types, such as Automobile, Finance, IT, etc. Each category contains thousands of documents. In this experiment, each category takes 400 documents, of which 280 are training set and 120 are test sets. Therefore, the training set contains 2520 documents, and the test set includes a total of 1080 documents.

The preprocessing module uses a third-party library for python, named jieba, to complete the work of word segmentation, part of speech tagging and discarding of stop words. In addition, we use Naive Bayesian classifier provided in Python's NLTK library as the classifier.

6.2.3.2 Experiments and Results

In this section, the following three groups of experiments are carried out to test the three innovation points of SM-CHI with control variable method. In Experiment I, we test the three feature selection algorithms without using synonym merges. In Experiment II, we use the first improved CHI method to select features and use grid search method to find the optimal threshold α. On the basis of Experiments I and II, we designed Experiment III to find the best weight update method.

Experiment I

In order to test the effect of three kinds of feature selection methods described in Sect. 6.2.1, we conducted the following experiments. But we didn't use synonym merging here. The results of experiment are shown in Fig. 6.10.

From the results, we can see that the two improved CHI formulas have a great effect on enhancing the value of F1 score of each category as compared to the original CHI formula, which means that the improved CHI formulas can select more representative words. They both make some improvement based on the original CHI. In addition, when the two improved CHI formulas are compared, the first improved method has a slight advantage, showing a better discrimination effect in the preceding categories. The result also shows that the log term successfully suppresses the CHI values of the high frequency words appearing in all classes, which achieves relatively good results. Therefore, we will use the first improved CHI formula as our base feature selection method in the fellow experiment.

Experiment II

In order to select the appropriate threshold α for synonymy merging, we designed the following experiment. The first improved CHI formula was used for feature selection, and the range of α is set to [0.5, 0.6, 0.7, 0.75, 0.8, 0.85, 0.9,1.0]. A total of nine experiments are conducted, including a comparative experiment that didn't use synonym merging. The experimental results are shown as Table 6.3 and Fig. 6.12.

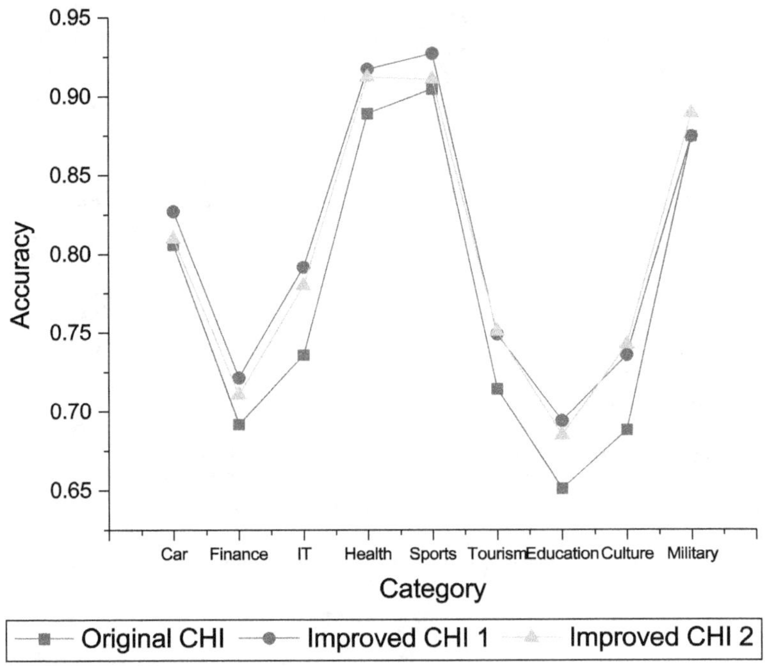

Fig. 6.10 Classification results comparing three kinds CHI formula

Table 6.3 F1 score of each category when α takes different values

	Car	Finance	IT	Health	Sports	Tourism	Edu	Culture	Mil
$\alpha = 0.50$	0.729	0.701	0.579	0.833	0.769	0.805	0.704	0.693	0.900
$\alpha = 0.60$	0.747	0.787	0.709	0.872	0.815	0.810	0.712	0.713	0.895
$\alpha = 0.70$	0.734	0.814	0.803	0.874	0.919	0.817	0.694	0.764	0.896
$\alpha = 0.75$	0.757	0.813	0.825	0.861	0.946	0.808	**0.726**	0.779	0.900
$\alpha = 0.80$	0.763	**0.821**	**0.828**	0.931	**0.949**	**0.827**	**0.726**	**0.819**	0.903
$\alpha = 0.85$	0.763	0.812	0.818	0.931	0.937	0.821	0.721	0.804	0.889
$\alpha = 0.90$	**0.782**	0.783	0.808	**0.943**	0.945	0.781	0.720	0.802	**0.913**
$\alpha = 1.0$	**0.782**	0.781	0.788	0.932	0.936	0.789	0.714	0.794	0.897
None	**0.782**	0.776	0.783	0.935	0.936	0.778	0.711	0.788	0.901

Bold values indicate the maximum value in each category

From the results in Figs. 6.11 and 6.12, we can draw the conclusion that the classification accuracy is the highest when $\alpha = 0.8$ and worst when $\alpha = 0.5$. The use of synonym merging improved classification accuracy by approximately 3 percentage points compared to use CHI only. By specific analysis of each category, we found that when we use synonym merging only the first category has a low accuracy compared to no synonym merging. The reason is that the eigenvectors after synonym merging have reduced the text discrimination degree of the "car" category.

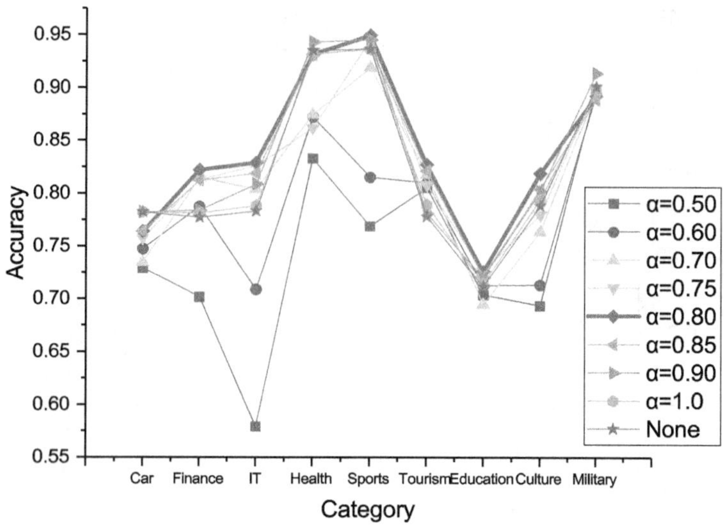

Fig. 6.11 F1 score of each category when α takes different values

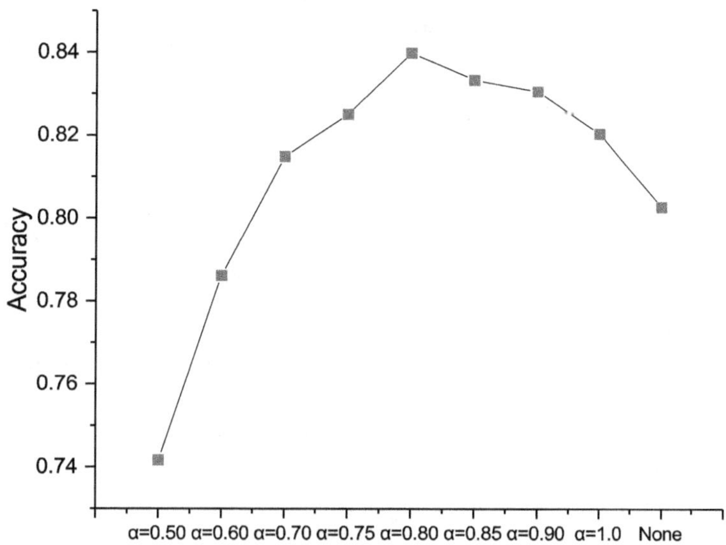

Fig. 6.12 Total accuracy when α takes different values

Experiment III

Based on the previous two experiments, we designed the following experiment to select the optimal weight update method introduced in Sect. 3.3. We designed three experiments, all with the first improved CHI formula and a threshold of 0.8. The experimental results are shown in Fig. 6.13:

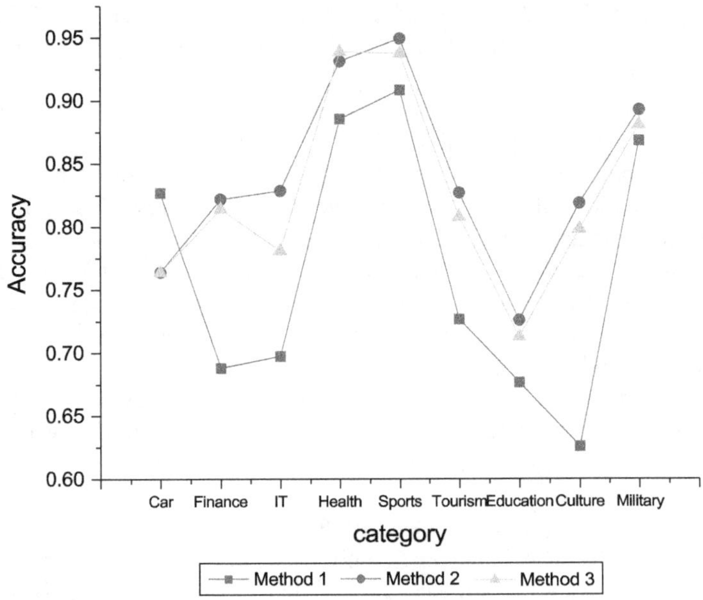

Fig. 6.13 Three kinds of special effects classification weight value updating method of comparison

According to Fig. 6.13, it can be seen that the classification accuracy of method 1 is the lowest. The reason is that this method adds the weights of all the synonymy words as the weight of the feature, but a word and its synonym words appear in more than one category. This simple superposition will make the feature differentiate the category worse. In contrast, methods 2 and 3 use the combined synonym as a one-dimensional feature and achieve better classification results and F1 scores. In contrast, method 2 is more effective which shows that the maximum value of the synonym is a better method because it can represent the maximum abilities of all synonyms to differentiate the text categories. By multiplying the power of 1.1 by the n, method 3 incorrectly increases the ability of the feature to distinguish text categories.

6.3 Sematic and Statistical Information Based Text Classification

In recent years, with an increasing volume of text information on the Internet and social media, text categorization has become a key technique to process these textual data. In text categorization, a Bag of Words (BOW) is usually used to represent a document. The weight in BOW is usually obtained by computing word frequency or the widely accepted TF-IDF formula. However, the BOW representation has several limitations: (1) The TF-IDF formula does not consider the calculation of class-based weights. As a result, the same word has the same weight in all categories. (2) It cannot deal with synonyms and polysemy.

In the absence of knowledge-based word similarity and statistic-based word similarity, automatic text categorization using BOW only as a document representation model[64] has not yet achieved the best performance and cannot meet the needs of all scenes in the real life. There are two ways to address this problem. First, we could use language model based on deep learning models such as word2vec [65] and Glove [29, 66] to learn the vector representations of words. However, such new approaches do not have to be necessarily better when the corpora is not particularly large. And it takes considerable time and effort to train word vectors. The second method is to collect semantic and syntactic informations as much as possible. We mainly adopt the second method and develop a new semantic smoothing kernel function based on knowledge-based word similarity and statistic-based word similarity to increase the capability of feature vectors to represent a document [33].

This chapter presents a novel approach for text classification. In this approach, word similarity based on HowNet is embedded into semantic information. This method promisingly improves the accuracy of text classification via using ontology knowledges. Moreover, the proposed approach takes advantage of the class-based term weighting by giving more weights on core words in each class during the transformation phase of SVM from the input space to the feature space. A term has a more discriminative power on a class if it has a higher weight for that class. The heuristic idea combining sematic and statistical information finally improves the classification accuracy.

6.3.1 Word Similarity

6.3.1.1 Support Vector Machines for Classification

Support vector machine (SVM) is a very effective machine learning algorithm developed from statistical learning theory. This algorithm was proposed by Vapnik, Guyon and Boser [67] and further analyzed in [68]. The core goal of SVM is to find the optimal segmentation hyperplane by the maximum spacing between classes. The author of [69] proposed that SVM has many advantages, such as finding the global optimal solution and having a good robustness.

SVM kernel function can be regarded as similarity function, because it calculates the similarity values of data sets. It is proposed to define a suitable kernel function [51, 70], which has a direct influence on finding the optimal hyperplane. The commonly used kernel functions for the document vectors are given below:

$$LinearKernel : k(d_p, d_q) = d_p d_q \tag{6.22}$$

$$PolynomialKernel : k(d_p, d_q) = (d_p d_q)^b, b = 1, 2, \ldots \tag{6.23}$$

$$RBFKernel : (d_p, d_q) = \exp(\gamma ||d_p - d_q||)^2 \tag{6.24}$$

In current works, the authors of [71] proposed to develop a kernel function based on the similarity of the knowledge system, and used the Omiotis library function to measure the similarity of English words and improve the accuracy of the classification. The authors of [72] proposed to develop a kernel function based on the weights of the class which improved the accuracy of the classifier. Based on these research efforts, this part optimizes the similarity of Chinese text words and combines the statistical methods with the knowledge-based methods to construct a kernel functions to improve the accuracy of text classification.

6.3.1.2 Knowledge-Based Word Similarity

Knowledge-based systems use ontology or thesaurus to capture the concepts in the documents and incorporate the domain knowledge into the words for the representation of textual data. These systems enhance the representation of terms by taking advantages of semantic relatedness among terms.

The similarity calculation of words is added to the text classification [34, 73], which is used to modify the weights of the text feature vectors. Mavroeidis et al. [74] proposed a semantic kernel function based on WordNet [75] to improve the accuracy of English text classification. Based on the Chinese semantic knowledge system of HowNet [76], Zhang embedded the semantic similarity into the kernel function of Chinese text classification [77], and improved the performance of Chinese text classification.

In this part, we use the Chinese dictionary HowNet to calculate the semantic similarity of words. HowNet is a very detailed dictionary of semantic knowledge. Unlike CiLin [78] and WordNet, every word in HowNet has multidimensional knowledge representations. The structure of HowNet is described in detail below.

HowNet mainly includes concepts and primitives. Each term is described by a number of concepts, each of which is described by a sequence of primitives, so primitive is the smallest expression unit in HowNet. HowNet contains 1500 primitives, which can be divided into three categories: basic semantics (describing the semantic features of concepts), grammatical semantics (describing the grammatical features of words) and relational semantics (describing the relationship between concepts). When we calculate the word similarity, we can define it in the following way. Word similarity calculation consists of four parts in Eq. (6.25).

$$Sim(S_1, S_2) = \sum_{i=1}^{4} \beta_i \prod_{j=1}^{i} Sim_j(S_1, S_2) \tag{6.25}$$

$Sim_1(S_1, S_2)$ is the similarity of first basic primitives of these two words. The similarity $Sim_1(S_1, S_2)$ between S_1 and S_2 can be calculated using Eq. (6.26). $Sim_2(S_1, S_2)$ is the similarity of the rest basic primitives, that is the arithmetic mean of the similarity of all pairs of elements. $Sim_3(S_1, S_2)$ is the similarity of two grammatical semantics, which can be transformed into the basic semantic meaning in the grammatical semantics. $Sim_4(S_1, S_2)$ is the similarity of two relational semantics, but the elements in the relational semantics are sets, which are basic primitives or concrete words.

There is a close relationship between word similarity and word distance. In fact, word similarity and word distance are different forms of the same feature of a pair of words. Word similarity is defined as a real number between 0 and 1.

$$Sim_1(S_1, S_2) = \frac{\alpha}{d + \alpha} \tag{6.26}$$

where S_1 and S_2 represent two of the words respectively, d is the distance between S_1 and S_2 in the original path hierarchy in HowNet; α is an adjustable parameter. When the distance in HowNet between words is particularly large, $Sim(S_1, S_2)$ approaches 0; when the distance in HowNet between words is particularly small, $Sim(S_1, S_2)$ approaches 1.

β_i in Eq. (6.25) is an adjustable parameters and satisfies Eq. (6.27). The latter part of Eq. (6.27) represents the descending importance of $Sim_1(S_1, S_2)$ to $Sim_4(S_1, S_2)$.

$$\beta_1 + \beta_2 + \beta_3 + \beta_4 = 1, \beta_1 \geq \beta_2 \geq \beta_3 \geq \beta_4 \tag{6.27}$$

6.3.1.3 Statistic-Based Word Similarity

In the absence of semantic knowledge, a statistical-based approach such as the one presented in [39, 79] can be applied to text categorization to solve synonymic problems. Statistical similarity calculation is based on the correlation of training words, so the statistical similarity calculation method is very sensitive to the training data sets.

The similarity calculation using statistics in text categorization contains calculation based on classes [72], high-order paths [80–82] and mean value calculations [83]. In this part, we use the method proposed in [72] which considers the weight of a certain word in a training set depending on the relevance of terms and categories. After the text feature vector is smoothed through the semantic kernel, it can improve the weights of important words in the category and reduce the weight of common words in the category. By modifying weights of the text feature vectors, the representation capability of the feature vectors is increased.

6.3.1.4 Weight of Feature Words

In the classification system, the most commonly used method of calculating word weights is the TF-IDF formula mentioned in [44, 58, 85, 86] where TF denotes the term frequency and IDF denotes inverse document frequency. TF-IDF formula was first used in the field of information retrieval, because its calculation method is simple and practical. It is also widely used in the text automatic classification.

TF-IDF is a statistical method to evaluate the importance of a document in a corpus. In general, the importance of a word increases in proportion to its number of occurrences in the document and decreases inversely with its higher frequency of occurrences in the corpus.

IDF formula is given in Eq. (6.28)

$$IDF(w) = \frac{|D|}{df_w} \tag{6.28}$$

where $|D|$ denotes the total number of documents in the corpus and df_w represents the number of documents which contains term w.

TF-IDF formula is given in Eq. (6.29)

$$TFIDF(w, d_i) = tf_w * \log(IDF(w)) \tag{6.29}$$

where tf_w represents the term frequency which is the number of word w in document D.

TF-ICF is proposed as another method of calculating word weight in [87, 88], which is similar to TF-IDF. ICF denotes inverse class frequency. TF-ICF calculates the word weight in category level rather than in document level.

Equation (6.30) shows the ICF formula:

$$ICF(w) = \frac{|C|}{cf_w} \tag{6.30}$$

where $|C|$ denotes the total number of classes in the corpus and cf_w represents the number of classes which contains term w.

TF-ICF formula is shown in Eq. (6.31),

$$TFICF(w, c_i) = \sum_{d \in c_j} tf_w * \log(ICF(w)) \tag{6.31}$$

Inspired by the IDF and ICF formulas, [48, 89, 90] proposes a new method for calculating weights:

$$W_{w,c} = \log(tfc_{w,c} + 1) * \frac{|D|}{df_w} \tag{6.32}$$

where $tfc_{w,c}$ represents the total number of feature term w of class c. $|D|$ denotes the total number of documents and df_w represents the number of documents which contain term w.

From analysis above, we can see that W is a matrix which is determined by categories and feature terms. In fact, terms that are similar to the topic in the category are given a larger weight because of the W matrix. The authors of [89, 90] compare the weighting algorithm based on the category with other commonly used feature selection algorithms, and conclude that the former one can improve the classification performance significantly.

6.3.2 *Merged Kernel Function*

Both knowledge-based word similarity computation and statistical-based word similarity computation can improve the performance of text classifiers [84]. A heuristic idea is to apply two computation methods to a kernel function which we call merged kernel function in this part. The pseudocode for the merged kernel function is shown in Algorithm 5. This section will explain in detail how to combine the two kernel functions to improve the accuracy of classification.

6.3.2.1 Vector Space Model

Document representation is a basic problem in natural language processing. Computer cannot directly deal with document which is mainly consisted of unstructured data. The key challenge is how to map a document into a vector space model (VSM). First, a document d_i is represented as an n-dimensional vector composed of feature words as shown in Eq. (6.33).

$$d_j = (w_{1j}, w_{2j}, \ldots, w_{nj}) \tag{6.33}$$

Then, the weighting formula maps the document vector d_i to the word weight vector $\phi(d_j)$:

$$\phi(d_j) = [tfidf(t_1, d_j), tfidf(t_2, d_j), \ldots, tfidf(t_n, d_j)] \tag{6.34}$$

where $tfidf(t_i, d_j)$ denotes the TF-IDF value of the feature word t_i in the document d_j.

6.3.2.2 Statistic-Based Similarity Matrix

The training phase for statistic-based similarity matrix is shown in Algorithm 5 [need to rewrite, just a figure] from the first line to the fourteenth line. In order

Algorithm 5 Calculation of the combined semantic smoothing matrix C

Require: Training set D
Ensure: Sematic Smoothing Matrix C
Local variables:
  $tfc_{\omega,k}$: total term frequency of word ω in the documents of class k
  $tf_{\omega,d}$: total term frequency of word ω in the document d
  N: total number of documents in the training set
  N_{ω}: shows total number of documents in the training set those contain word w
  Z: matrix is constructed according to the list of feature word by the introduction of HowNet
for each word w **do**
  for each document d contains word w in the train set **do**
    $N_{\omega} = N_{\omega} + 1$
  end for
end for

for each word w **do**
  for each document d contains word w in class k **do**
    $tfc_{\omega,k} = tfc_{\omega,k} + tf_{\omega,d}$
  end for
  $W_{\omega,k} = (\log(tfc_{\omega,k}) + 1) * \log(N/N_{\omega})$
end for

$S = W * W^T$

$Z^2 = Z * Z^T$

for i in columns of S **do**
  for j in rows of S **do**
    $C_{i,j} = \lambda_1 * S_{ij} + \lambda_2 * Z^2_{ij}$
  end for
end for

to embed the statistical information into the space vector model and increase the capability of representing a document of feature vectors, we construct a matrix S based on the class-based weight, which is called the statistical similarity matrix. The formula has been described in detail in Sect. 6.3.1.4. To make use of this formula, we define the statistical similarity matrix S as:

$$S = WW^T \tag{6.35}$$

where W is the class-based weighting formula mentioned previously. The statistical similarity matrix S is a symmetric matrix, and the element S_{ij} in matrix S is the class-based statistical similarity of the feature words w_i and w_j.

The S matrix represents the statistical similarity of words. For example, the words “patient” and “sufferer” have similar meaning. When the weight of the “patient” is higher and the weight of the “sufferer” is lower in vector $\varphi(d_j)$, the weight of the word sufferer will be increased after multiplication with the S matrix.

6.3.2.3 Combined Semantic Smoothing Matrices

The knowledge-based similarity matrix Z can be constructed according to the list of feature words by the introduction of HowNet. Z_{ij} denotes the knowledge-based similarity between the feature words w_i and w_j. The knowledge-based word similarity can be embedded into the space vector model by matrix Z. In order to ensure the validity of final merged kernel, a second-order rule similarity matrix is used here, that is $Z^2 = ZZ^T$.

After computing the statistic-based similarity matrix S and the second-order knowledge-based similarity matrix Z^2, the degree of weight modification of the two matrices to the space vector model cannot be determined, which should be determined according to the data set. In this part, the matrix S and Z^2 are combined to generate a new semantic smoothing matrix C:

$$C_{ij} = \lambda_1 * S_{ij} + \lambda_2 * Z_{ij}^2 \tag{6.36}$$

where S_{ij} and Z_{ij}^2 are described in Sects. 6.3.2.2 and 6.3.2.3, λ_1 and λ_2 adjust the normalization parameters of the weights in S and Z^2. Parameters λ_1 and λ_2 satisfy $\lambda_1 + \lambda_2 = 1$. We can adjust λ_1 and λ_2 to determine how matrices S and Z^2 affect the classification performance of the classifier.

6.3.2.4 Semantic Smoothing Kernel Function

By defining the mapping architecture matrix C, we can map a document vector to a new feature space vector by using Eq. (6.37).

$$\overline{\phi}(d_j) = \phi(d_j)C \tag{6.37}$$

The mapped vectors can be directly used in many classification methods. If the high-dimensional sparse matrix appears in the text classification, there will be a high-dimensional disaster in computation. Defining a kernel function can reduce the influence of the high-dimensional sparse matrix. The inner product between documents p and q in the feature space is computed by the kernel function using Eq. (6.38).

$$K_{CK}(d_p, d_q) =< \overline{\phi}(d_p), \overline{\phi}(d_q) >= \phi(d_p)CC^T\phi(d_q)^T \tag{6.38}$$

where $K_{CK}(d_p, d_q)$ denotes the similarity of document d_p and document d_q. $\overline{\phi}(d_p)$ and $\overline{\phi}(d_q)$ are the new feature space vectors of document d_p and document d_q after transformed by the semantic smoothing matrix proposed in Eq. (6.37). The kernel function information is stored in the matrix G:

$$G_{p,q} = K_{CK}(d_p, d_q) \tag{6.39}$$

Then we prove the validity of the semantic smooth kernel proposed in this section. According to mercer theorem [91], any semi-definite function can be used as a kernel function. The semantic smoothing matrix C proposed in this part is composed of the statistical similarity matrix S and the second-order knowledge-based similarity matrix Z^2, so the matrix C is also a symmetric matrix. The matrix S and the matrix Z^2 are the product of a matrix and its transpositions, so the matrix S and Z^2 are both semi-definite matrix, which is proved in [92]. In linear algebra, the sum of two positive semi-definite matrices is also a semi-definite matrix. Therefore, the matrix C is a semi-definite matrix, satisfying the conditions required by Mercer's theorem. The kernel function can be constructed by the semantic smoothing matrix C.

6.3.3 *Experiments and Results*

6.3.3.1 Corpora

This section selects the corpus provided by Sogou Company.[1] and Fudan University.[2] The Sogou corpus consists of SogouCA and SogouCS news corpora containing various categories of 2,909,551 news articles, of which about 2,644,110 articles contain both a title and relevant content. We manually categorize articles by using the channel in URL, then we get a large Chinese corpus with the article contents and categorizes. However, there are some categorizes that contain few articles. So five categories with largest number—"sports", "finance", "entertainment", "automobile" and "technology" are finally selected for our text classification experiments. The details of the Sogou corpus are presented in Table 6.4. During the training process, many parameters in the model are involved. In order to determine the optimal values of parameters in this model, we use the validation set. We partition training set, validation set and test set in 8:1:1 proportions in Sogou corpora after we shuffled the corpora. So the corpora is randomly divided into training set, validation set and test set. These parameters are described in detail in Sect. 6.3.3.4.

Table 6.4 Sogou news corpora

Category	Total	Train	Validation	Test
Sports	645,931	80,000	10,000	10,000
Finance	315,551	80,000	10,000	10,000
Entertainment	160,409	80,000	10,000	10,000
Automobile	167,647	80,000	10,000	10,000
Technology	188,111	80,000	10,000	10,000

[1]http://www.sogou.com/labs/resource/list_news.php.

[2]http://www.nlpir.org/download/tc-corpus-answer.rar.

Table 6.5 Fudan University corpora

Category	Total	Train	Validation	Test
Economy	1589	700	100	100
Sports	1188	700	100	100
Environment	1022	700	100	100
Politics	1013	700	100	100
Agriculture	992	700	100	100

To validate the combine kernel's effect on a small corpora, we also use the corpus provided by Fudan University. The corpora contains 9804 articles that have been already divided into 20 categories. We choose five categories—"economy", "sports", "environment", "politics" and "agriculture". The details of the Fudan training set are presented in Table 6.5. We partition training set, validation set and test set in 7:1:1 proportions in Fudan corpus after we shuffled the corpora.

6.3.3.2 Word Segmentation and Stop Words

The current English word segmentation tool has been well developed, while the Chinese word segmentation technology is still evolving. For python language, NLTK [93] nltk.tokenize module can be used for word segmentation in English, and jieba tool can be used for word segmentation in Chinese.

In order to save storage and improve the efficiency of classification, the classification system will ignore certain words after word classification, which are called stop words.[3] There are two kinds of stop words: the first one can be found everywhere in all kinds of documents with which the classification system cannot guarantee the true classification result. The second kind of stop words includes the modal particle, adverb, preposition, conjunction and so on.

6.3.3.3 Feature Word Extraction

In the problem of text categorization, a certain feature word and its class obey the CHI square distribution. The larger the CHI value is, the more the CHI value can be used to identify the category. The CHI formula is given:

$$\chi^2(t, c) = \frac{N(AD - BC)^2}{(A + C)(A + B)(B + D)(C + D)} \tag{6.40}$$

where N is the number of texts in the training set, A is the number of documents belonging to class c and contains the word w; B is the number of documents that do not belong to class c but contains word w; C is the number of documents belonging

[3] https://github.com/Irvinglove/Chinese_stop_words/blob/master/stopwords.txt.

to c Class, but does not contain the word w; D is the number of documents that do not belong to c class and do not contain the word w.

6.3.3.4 Experiment Settings

The classifier uses the SVM function provided in the machine learning library sklearn [94] in python environment. We change the kernel function by using the interface it provides. We observe how the statistical similarity matrix S and the second-order knowledge-based similarity matrix Z^2 affect the performance of the classifier when the ratio of training set and parameter λ are different.

In the experiment, we set parameter values based on the experience gained from the validation set. First, we describe several parameters mentioned above. We compute word similarity using α 1.6, β_1 0.5, β_2 0.2, β_3 0.17 and β_4 0.13. Second, the length of VSM used for representing Sogou corpus is 10,000, and Fudan corpus is 1000. Sogou corpus is large, so the feature vectors need to be longer. Finally, we set some parameters of the model. Penalty parameter C of the error term is 1.0 and there is no hard limit on iterations within solver, so that the SVM algorithm will stop training before over-fitting.

A number of parameters have been used to assess the performance of classification model output, such as accuracy [95] and f-measure (F1) [96]. To demonstrate that the combined kernel does improve the accuracy and F1 value of text classification, we use other machine learning methods for comparison, including KNN, Naive Bayes, and SVM with linear kernel and RBF kernel. The corpora is processed in the same way in Sects. 6.3.3.2 and 6.3.3.3, and then we call the interface of different machine learning algorithms in sklearn. We compare the results with character-level convolutional [97] networks which is the state-of-the-art method in text classification. Finally we adopt the accuracy and the F1 value of text classification as the evaluation standard.

6.3.3.5 Experiments and Results

As shown in Tables 6.6 and 6.7, the first column in the table shows the compared training algorithms, and the first row indicates the value of λ_1. The value of λ_2 corresponding to this is 1-λ_1. The second row shows the performance of the combined kernel in the case of different λ_1 value. The rest rows represent the performance of other machine learning methods. The values in Table 6.6 represent the accuracy rate of Sogou corpus and the values in Table 6.7 represent the values of F1 of Sogou corpus.

The values obtained in Tables 6.6 and 6.7 are shown by the line chart in Figs. 6.14 and 6.15 respectively from which it is easier to see the effect of the combination of the statistical similarity matrix S and the second-order knowledge-based similarity matrix Z^2 on the classification accuracy.

Table 6.6 Accuracy rate of Sogou corpus

	0	0.2	0.4	0.6	0.8	1
SVM (Combined kernel)	92.12%	94.98%	95.58%	96.43%	95.32%	93.12%
CNN (Character-level)	95.12%					
SVM (Linear kernel)	75.23%					
SVM (RBF kernel)	82.53%					
KNN	78.62%					
Bayes	85.65%					

Table 6.7 F1 value of Sogou corpus

	0	0.2	0.4	0.6	0.8	1
SVM (Combined kernel)	92.16%	93.68%	94.96%	95.75%	95.14%	93.36%
CNN (Character-level)	94.93%					
SVM (Linear kernel)	77.32%					
SVM (RBF kernel)	84.52%					
KNN	78.53%					
Bayes	85.12%					

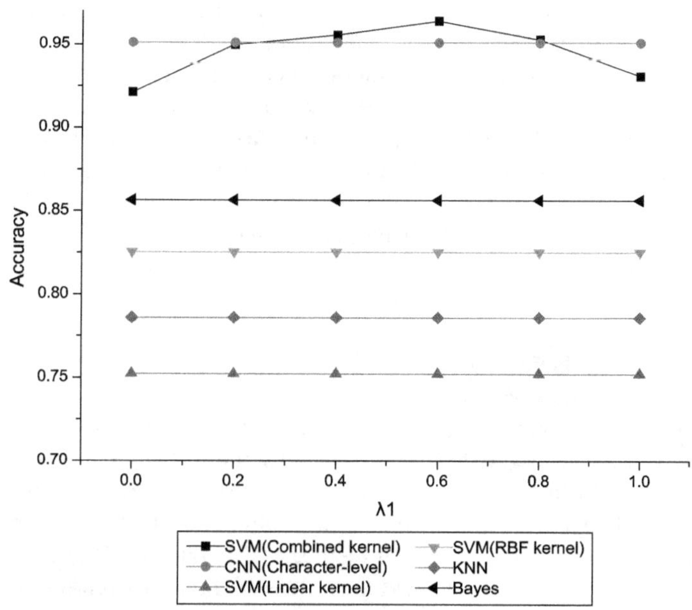

Fig. 6.14 Curve: accuracy rate of Sogou corpus

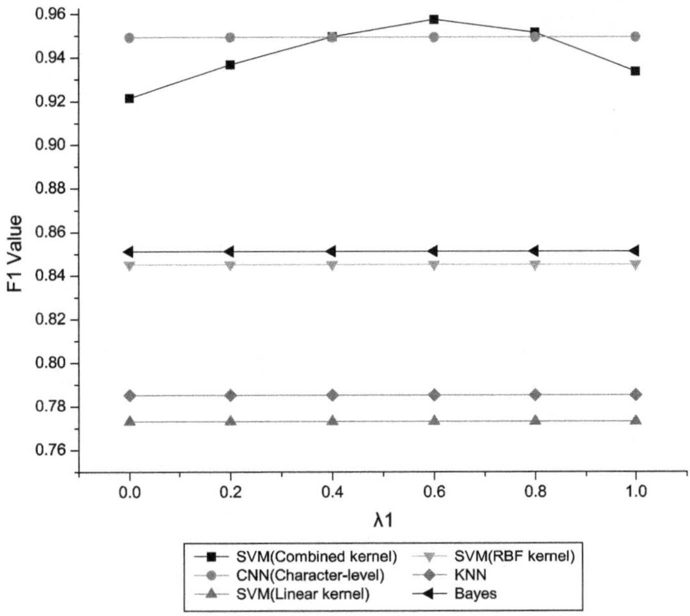

Fig. 6.15 Curve: F1 value of Sogou corpus

Table 6.8 Accuracy rate of Fudan corpus

	0	0.2	0.4	0.6	0.8	1
SVM (Combined kernel)	95.34%	95.75%	96.68%	96.23%	95.56%	95.14%
CNN (Character-level)	95.15%					
SVM (Linear kernel)	89.41%					
SVM (RBF kernel)	94.32%					
KNN	90.21%					
Bayes	95.59%					

As shown in Fig. 6.14, the accuracy rate is lower than that using character-level convolutional networks which is a very effective classification method in text categorization when λ_1 is 0, 0.2, and 1. However, the accuracy rate can be maintained at a high level when λ_1 is between 0.4 and 0.8. And the accuracy rate is always higher than that using KNN, Bayes and SVM with linear kernel and RBF kernel, proving the combination of the two is meaningful for Chinese text classification.

The values in Table 6.8 represent the accuracy rate of Fudan corpus and the values in Table 6.9 represent the values of F1 of Fudan corpus. The values obtained in Tables 6.8 and 6.9 are shown by the line chart in Figs. 6.16 and 6.17, from which we can confirm that the combination of the two kernels is meaningful for Chinese text classification.

Table 6.9 F1 value of Fudan corpus

	0	0.2	0.4	0.6	0.8	1
SVM (Combined kernel)	96.12%	96.41%	96.15%	97.24%	96.58%	96.07%
CNN (Character-level)	95.37%					
SVM (Linear kernel)	90.24%					
SVM (RBF kernel)	94.84%					
KNN	89.56%					
Bayes	96.32%					

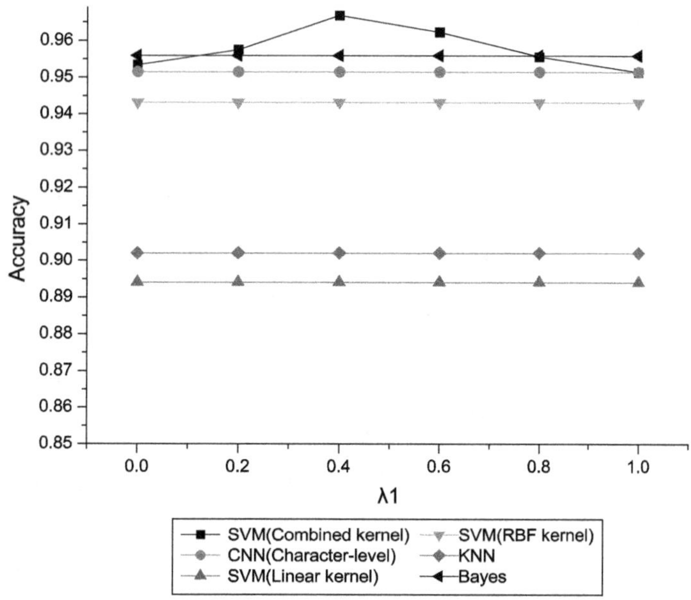

Fig. 6.16 Curve: accuracy rate of Fudan corpus

6.4 Summary

In this chapter, we discuss the main challenge of intend-based networking and introduced several algorithms. We first propose an effective model for the similarity metrics of English sentences. Then, the SM-CHI feature selection method based on the common method used in Chinese text classification is presented. Finally, we proposed a novel approach which considers both the semantic and statistical information to improve the accuracy of text classification.

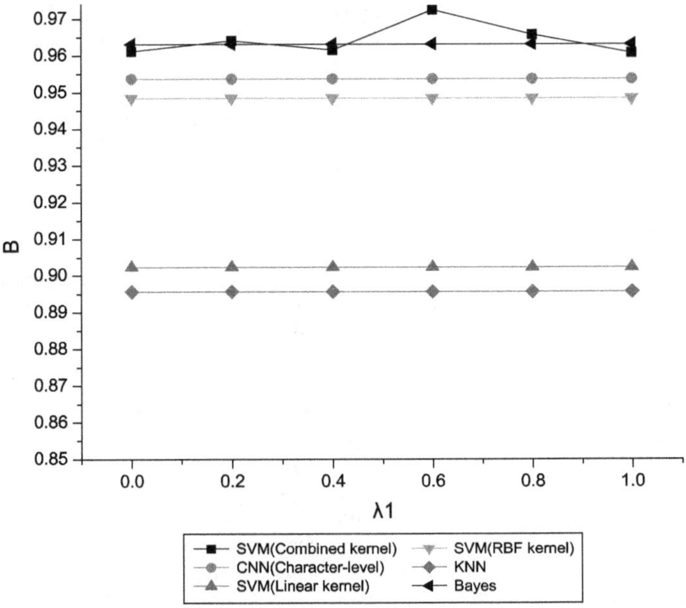

Fig. 6.17 Curve: F1 value of Fudan corpus

References

1. Mikolov Tomas, Sutskever Ilya, Chen Kai, Corrado Greg S, Dean Jeff. Distributed Representations of Words and Phrases and their Compositionality. In: Burges C. J. C., Bottou L., Welling M., Ghahramani Z., Weinberger K. Q., eds. *Advances in Neural Information Processing Systems 26*, Curran Associates, Inc. 2013 (pp. 3111–3119).
2. C. Jiang, H. Zhang, R. Yong, and H. H. Chen, "Energy-efficient non-cooperative cognitive radio networks: Micro, meso, and macro views," *IEEE Communications Magazine*, vol. 52, no. 7, pp. 14–20, 2014.
3. Mikolov Tomas, Chen Kai, Corrado Greg, Dean Jeffrey. Efficient Estimation of Word Representations in Vector Space. *Computation and Language.* 2013.
4. S. Lei, L. Zhou, Q. Peng, and H. Yao, "Openflow based spatial information network architecture," in *International Conference on Wireless Communications & Signal Processing*, 2015.
5. Collobert Ronan, Weston Jason. A unified architecture for natural language processing: deep neural networks with multitask learning. In: International conference on machine learning:160–167; 2008.
6. Hu Baotian, Lu Zhengdong, Li Hang, Chen Qingcai. Convolutional neural network architectures for matching natural language sentences. In: International Conference on Neural Information Processing Systems:2042–2050; 2014.
7. Yin Wenpeng, Schütze Hinrich. Convolutional Neural Network for Paraphrase Identification. In: Conference of the North American Chapter of the Association for Computational Linguistics: Human Language Technologies:901–911; 2015.
8. C. Jiang, C. Yan, and K. J. R. Liu, "Graphical evolutionary game for information diffusion over social networks," *IEEE Journal of Selected Topics in Signal Processing*, vol. 8, no. 4, pp. 524–536, 2017.

9. wikipedia. tf-idf. https://en.wikipedia.org/wiki/Tf%E2%80%93idf.
10. Tai Kai Sheng, Socher Richard, Manning Christopher D. Improved Semantic Representations From Tree-Structured Long Short-Term Memory Networks. *Computer Science.* 2015;5(1): 36.
11. F. Xu, L. Rong, C. Zhao, H. Yao, and J. Zhang, "Congestion-aware signaling aggregation scheme for cellular based underwater acoustic sensor network," in *IEEE International Conference on Communication Workshop*, 2015.
12. Kalchbrenner Nal, Grefenstette Edward, Blunsom Phil. A Convolutional Neural Network for Modelling Sentences. *Meeting of the association for computational linguistics.* 2014;655–665.
13. Kim Yoon. Convolutional Neural Networks for Sentence Classification. *Empirical methods in natural language processing.* 2014;1746–1751.
14. He Hua, Gimpel Kevin, Lin Jimmy. Multi-Perspective Sentence Similarity Modeling with Convolutional Neural Networks. In: Conference on Empirical Methods in Natural Language Processing:1576–1586; 2015.
15. ——, "Multi-channel sensing and access game: Bayesian social learning with negative network externality," *IEEE Transactions on Wireless Communications*, vol. 13, no. 4, pp. 2176–2188, 2014.
16. He Hua, Lin Jimmy. Pairwise Word Interaction Modeling with Deep Neural Networks for Semantic Similarity Measurement. In: Conference of the North American Chapter of the Association for Computational Linguistics: Human Language Technologies:937–948; 2016.
17. H. Yao, X. Sun, Z. Zhou, L. Tang, and L. Shi, "Joint optimization of subchannel selection and spectrum sensing time for multiband cognitive radio networks," in *International Symposium on Communications & Information Technologies*, 2010.
18. S. Lei, Z. Zheng, T. Liang, H. Yao, and Z. Jing, "Ultra-wideband channel estimation based on Bayesian compressive sensing," in *International Symposium on Communications & Information Technologies*, 2010.
19. Socher Richard, Karpathy Andrej, Le Quoc V., Manning Chris D., Ng Andrew Y.. Grounded compositional semantics for finding and describing images with sentences. *Nlp.stanford.edu.* 2013.
20. Mueller Jonas, Thyagarajan Aditya. Siamese recurrent architectures for learning sentence similarity. In: Thirtieth AAAI Conference on Artificial Intelligence:2786–2792; 2016.
21. C. Jiang, Y. Chen, Y. H. Yang, C. Y. Wang, and K. J. R. Liu, "Dynamic Chinese restaurant game: Theory and application to cognitive radio networks," *IEEE Transactions on Wireless Communications*, vol. 13, no. 4, pp. 1960–1973, 2014.
22. T. Liang, Z. Zhou, L. Shi, H. Yao, J. Zhang, and Y. Ye, "Laplace prior based distributed compressive sensing," in *International 1st Conference on Communications & Networking in China*, 2010.
23. Erk Katrin. A structured vector space model for word meaning in context. In: Conference on Empirical Methods in Natural Language Processing:897–906; 2008.
24. Z. Zhen, T. Jiang, W. S. Zhang, and H. Yao, "Analysis speech of polypus patients based on channel parameters and fuzzy logic systems," in *Seventh International Conference on Fuzzy Systems & Knowledge Discovery*, 2010.
25. Lapata Mirella, Mitchell Jeff. Vector-based Models of Semantic Composition. 2008.
26. Mitchell J, Lapata M. Composition in distributional models of semantics. *Cognitive Science.* 2010;34(8):1388.
27. Turney Peter D. Domain and Function: A Dual-Space Model of Semantic Relations and Compositions. *Journal of Artificial Intelligence Research.* 2013;44(4):533–585.
28. Y. H. Yang, Y. Chen, C. Jiang, C. Y. Wang, and K. J. R. Liu, "Wireless access network selection game with negative network externality," *IEEE Transactions on Wireless Communications*, vol. 12, no. 10, pp. 5048–5060, 2013.
29. H. Yao, Z. Zheng, L. He, and L. Zhang, "Optimal power allocation in joint spectrum underlay and overlay cognitive radio networks," in *International Conference on Cognitive Radio Oriented Wireless Networks & Communications*, 2009.

30. Erk Katrin. Vector Space Models of Word Meaning and Phrase Meaning: A Survey. *Language & Linguistics Compass.* 2012;6(10):635–653.
31. Clarke Daoud. A Context-theoretic Framework for Compositionality in Distributional Semantics. *Computational Linguistics.* 2011;38(1):41–71.
32. Das Dipanjan, Smith Noah A. Paraphrase identification as probabilistic quasi-synchronous recognition. In: Joint Conference of the Meeting of the ACL and the International Joint Conference on Natural Language Processing of the Afnlp: Volume:468–476; 2009.
33. C. Jiang, C. Yan, and K. J. R. Liu, "Distributed adaptive networks: A graphical evolutionary game-theoretic view," *IEEE Transactions on Signal Processing*, vol. 61, no. 22, pp. 5675–5688, 2013.
34. H. Yao, Z. Zheng, L. Zhang, L. He, T. Liang, and K. S. Kwak, "An efficient power allocation scheme in joint spectrum overlay and underlay cognitive radio networks," in *International Conference on Communications & Information Technologies*, 2009.
35. Dolan Bill, Quirk Chris, Brockett Chris. Unsupervised construction of large paraphrase corpora: exploiting massively parallel news sources. In: International Conference on Computational Linguistics:350; 2004.
36. Nakov Preslav, Zesch Torsten, eds. Proceedings of the 8th International Workshop on Semantic Evaluation. The Association for Computer Linguistics 2014.
37. Marelli Marco, Bentivogli Luisa, Baroni Marco, Bernardi Raffaella, Menini Stefano, Zamparelli Roberto. Semeval-2014 Task 1: Evaluation of compositional distributional semantic models on full sentences through semantic relatedness and textual entailment. In: Semeval 2014: International Workshop on Semantic Evaluation:16; 2014.
38. Zhao Jiang, Zhu Tiantian, Lan Man. ECNU: One Stone Two Birds: Ensemble of Heterogenous Measures for Semantic Relatedness and Textual Entailment. In: International Workshop on Semantic Evaluation:271–277; 2014.
39. B. Li, Z. Zheng, W. Zou, K. S. Kwak, F. Wu, and H. Yao, "A nonlinear transform and its application in the optimum receiving of ultra narrow-band," in *International Conference on Communications & Information Technologies*, 2009.
40. Bjerva Johannes, Bos Johan, Goot Rob Van Der, Nissim Malvina. The Meaning Factory: Formal Semantics for Recognizing Textual Entailment and Determining Semantic Similarity. In: SemEval-2014 Workshop; 2014.
41. Jimenez Sergio, Dueñas George Enrique, Baquero Julia, Gelbukh Alexander. UNAL-NLP: Combining Soft Cardinality Features for Semantic Textual Similarity, Relatedness and Entailment. In: Semeval; 2014.
42. Lai Alice, Hockenmaier Julia. Illinois-LH: A Denotational and Distributional Approach to Semantics. In: International Workshop on Semantic Evaluation:329–334; 2014.
43. Bestgen Yves. CECL: a New Baseline and a Non-Compositional Approach for the Sick Benchmark. In: International Workshop on Semantic Evaluation:160–165; 2014.
44. H. Yao, L. Zhang, Z. Ran, and Z. Zheng, "An efficient game-based competitive spectrum offering scheme in cognitive radio networks with dynamic topology," in *IEEE International Conference on Communication Technology*, 2008.
45. Proisl Thomas, Evert Stefan, Greiner Paul, Kabashi Besim. SemantiKLUE: Robust Semantic Similarity at Multiple Levels Using Maximum Weight Matching. In: International Workshop on Semantic Evaluation:532–540; 2014.
46. Rus Vasile, Mccarthy Philip M., Lintean Mihai C., Mcnamara Danielle S., Graesser Arthur C. Paraphrase Identification with Lexico-Syntactic Graph Subsumption. In: International Florida Artificial Intelligence Research Society Conference, May 15–17, 2008, Coconut Grove, Florida, USA:201–206; 2008.
47. Blacoe William, Lapata Mirella. A comparison of vector-based representations for semantic composition. In: Joint Conference on Empirical Methods in Natural Language Processing and Computational Natural Language Learning:546–556; 2012.
48. Z. Ran, L. Zhang, Y. Fei, H. Yao, and Z. Zheng, *Balanced clustering multi-hop routing algorithm for LEACH protocol in wireless sensor networks*, 2008.

49. Fernando Samuel, Stevenson Mark. A Semantic Similarity Approach to Paraphrase Detection. *Computational Linguistics UK Annual Research Colloquium.* 2008.
50. Yao Haipeng, Liu Chong, Zhang Peiying, Wang Luyao. A feature selection method based on synonym merging in text classification system. *Eurasip Journal on Wireless Communications & Networking.* 2017;2017(1):166.
51. C. Jiang, C. Yan, G. Yang, and K. J. R. Liu, "Joint spectrum sensing and access evolutionary game in cognitive radio networks," *IEEE Transactions on Wireless Communications*, vol. 12, no. 5, pp. 2470–2483, 2013.
52. Hwee Tou Ng, Wei Boon Goh, and Kok Leong Low. Feature selection, perceptron learning, and a usability case study for text categorization. In *International ACM SIGIR Conference on Research and Development in Information Retrieval*, pages 67–73. ACM, 1997.
53. Yiming Yang and Jan O Pedersen. A comparative study on feature selection in text categorization. In *Fourteenth International Conference on Machine Learning*, pages 412–420. Morgan Kaufmann Publishers Inc, 1997.
54. Gui Chuan Feng and Shubin Cai. An improved feature extraction algorithm based on chi and mi. 2015.
55. Yan Tang and Ting Xiao. An improved χ^2 (chi) statistics method for text feature selection. In *International Conference on Computational Intelligence and Software Engineering*, pages 1–4. IEEE, 2009.
56. Thorsten Joachims. *Text categorization with Support Vector Machines: Learning with many relevant features*. Springer Berlin Heidelberg, 1998.
57. H. P. Yao, L. Y. Zhang, Z. Zhou, and X. U. Fang-Min, "A handoff algorithm based on grey prediction model for bluetooth network," *Radio Engineering of China*, 2008.
58. C. Jiang, Y. Chen, K. J. R. Liu, and Y. Ren, "Renewal-theoretical dynamic spectrum access in cognitive radio networks with unknown primary behavior," *IEEE Journal on Selected Areas in Communications*, vol. 31, no. 3, pp. 406–416, 2013.
59. Ted Dunning. Accurate methods for the statistics of surprise and coincidence. *Linguistics-74 Computational Dirk Geeraerts Stefan Grondelaers and*, 19(1):61–74, 1993.
60. J. Tian and W. Zhao. Words similarity algorithm based on tongyici cilin in semantic web adaptive learning system. *Journal of Jilin University*, 28(06):602–608, 2010.
61. H. Yao, Z. Zhang, and Y. Liu, "Research on the embedded sim technology in internet of things," *Information & Communications Technologies*, 2012.
62. Sijun Qin, Jia Song, Pengzhou Zhang, and Yue Tan. Feature selection for text classification based on part of speech filter and synonym merge. In *International Conference on Fuzzy Systems and Knowledge Discovery*, pages 681–685, 2015.
63. H. Yao, Z. Yang, H. Jiang, and L. Ma, "A scheme of ad-hoc-based d2d communication in cellular networks." *Adhoc & Sensor Wireless Networks*, vol. 32, 2016.
64. Susan Dumais, John Platt, David Heckerman, and Mehran Sahami. Inductive learning algorithms and representations for text categorization. In *Proceedings of the seventh international conference on Information and knowledge management*, pages 148–155. ACM, 1998.
65. Quoc V. Le and Tomas Mikolov. Distributed representations of sentences and documents. 4:II–1188, 2014.
66. Jeffrey Pennington, Richard Socher, and Christopher Manning. Glove: Global vectors for word representation. In *Conference on Empirical Methods in Natural Language Processing*, pages 1532–1543, 2014.
67. Bernhard E Boser, Isabelle M Guyon, and Vladimir N Vapnik. A training algorithm for optimal margin classifiers. In *Proceedings of the fifth annual workshop on Computational learning theory*, pages 144–152. ACM, 1992.
68. Stephan R Sain. The nature of statistical learning theory. *Technometrics*, 38(4):409–409, 1996.
69. Thorsten Joachims. Text categorization with support vector machines: Learning with many relevant features. In *European conference on machine learning*, pages 137–142. Springer, 1998.

70. Shun-ichi Amari and Si Wu. Improving support vector machine classifiers by modifying kernel functions. *Neural Networks*, 12(6):783–789, 1999.
71. Jamal Abdul Nasir, Asim Karim, George Tsatsaronis, and Iraklis Varlamis. A knowledge-based semantic kernel for text classification. In *International Symposium on String Processing and Information Retrieval*, pages 261–266. Springer, 2011.
72. Berna Altınel, Banu Diri, and Murat Can Ganiz. A novel semantic smoothing kernel for text classification with class-based weighting. *Knowledge-Based Systems*, 89:265–277, 2015.
73. George Siolas and Florence d'Alché Buc. Support vector machines based on a semantic kernel for text categorization. In *Neural Networks, 2000. IJCNN 2000, Proceedings of the IEEE-INNS-ENNS International Joint Conference on*, volume 5, pages 205–209. IEEE, 2000.
74. Dimitrios Mavroeidis, George Tsatsaronis, Michalis Vazirgiannis, Martin Theobald, and Gerhard Weikum. Word sense disambiguation for exploiting hierarchical thesauri in text classification. In *European Conference on Principles of Data Mining and Knowledge Discovery*, pages 181–192. Springer, 2005.
75. C Fellbaum and G Miller. *WordNet:An Electronic Lexical Database*. MIT Press, 1998.
76. Yan-Lan Zhu, Jin Min, Ya-qian Zhou, Xuan-jing Huang, and Li-De Wu. Semantic orientation computing based on hownet. *Journal of Chinese Information Processing*, 20(1):14–20, 2006.
77. Pei-Ying Zhang. A hownet-based semantic relatedness kernel for text classification. *Indonesian Journal of Electrical Engineering and Computer Science*, 11(4):1909–1915, 2013.
78. Jiaju Mei. *Tongyi ci cilin*. Shangai cishu chubanshe, 1985.
79. Nicholas E. Evangelopoulos. Latent semantic analysis. *Wiley interdisciplinary reviews. Cognitive science*, 4(6):683, 2013.
80. Berna Altlnel, Murat Can Ganiz, and Banu Diri. A novel higher-order semantic kernel for text classification. In *Electronics, Computer and Computation (ICECCO), 2013 International Conference on*, pages 216–219. IEEE, 2013.
81. Berna Altinel, Murat Can Ganiz, and Banu Diri. A semantic kernel for text classification based on iterative higher-order relations between words and documents. In *International Conference on Artificial Intelligence and Soft Computing*, pages 505–517. Springer, 2014.
82. Berna Altinel, Murat Can Ganiz, and Banu Diri. A simple semantic kernel approach for svm using higher-order paths. In *Innovations in Intelligent Systems and Applications (INISTA) Proceedings, 2014 IEEE International Symposium on*, pages 431–435. IEEE, 2014.
83. Berna Altınel, Murat Can Ganiz, and Banu Diri. A corpus-based semantic kernel for text classification by using meaning values of terms. *Engineering Applications of Artificial Intelligence*, 43:54–66, 2015.
84. H. Yao, Y. Liu, and C. Fang, "An abnormal network traffic detection algorithm based on big data analysis." *International Journal of Computers, Communications & Control*, vol. 11, no. 4, 2016.
85. Karen Sparck Jones. A statistical interpretation of term specificity and its application in retrieval. *Journal of documentation*, 28(1):11–21, 1972.
86. Gerard Salton and Christopher Buckley. Term-weighting approaches in automatic text retrieval. *Information processing & management*, 24(5):513–523, 1988.
87. Youngjoong Ko and Jungyun Seo. Automatic text categorization by unsupervised learning. In *Proceedings of the 18th conference on Computational linguistics-Volume 1*, pages 453–459. Association for Computational Linguistics, 2000.
88. Verayuth Lertnattee and Thanaruk Theeramunkong. Analysis of inverse class frequency in centroid-based text classification. In *Communications and Information Technology, 2004. ISCIT 2004. IEEE International Symposium on*, volume 2, pages 1171–1176. IEEE, 2004.
89. Göksel Biricik, Banu Diri, Ahmet Co, et al. A new method for attribute extraction with application on text classification. In *Soft Computing, Computing with Words and Perceptions in System Analysis, Decision and Control, 2009. ICSCCW 2009. Fifth International Conference on*, pages 1–4. IEEE, 2009.
90. GÖKSEL BİRİCİK, Banu Diri, and AHMET COŞKUN SÖNMEZ. Abstract feature extraction for text classification. *Turkish Journal of Electrical Engineering & Computer Sciences*, 20(Sup. 1):1137–1159, 2012.

91. Simon Parsons. Introduction to machine learning by ethem alpaydin, MIT press, 0-262-01211-1, 400 pp., $50.00/£ 32.95, 2005.
92. Nello Cristianini, John Shawe-Taylor, and Huma Lodhi. Latent semantic kernels. *Journal of Intelligent Information Systems*, 18(2-3):127–152, 2002.
93. Edward Loper and Steven Bird. Nltk: the natural language toolkit. In *Acl-02 Workshop on Effective TOOLS and Methodologies for Teaching Natural Language Processing and Computational Linguistics*, pages 63–70, 2002.
94. Fabian Pedregosa, Alexandre Gramfort, Vincent Michel, Bertrand Thirion, Olivier Grisel, Mathieu Blondel, Peter Prettenhofer, Ron Weiss, Vincent Dubourg, and Jake Vanderplas. Scikit-learn: Machine learning in python. *Journal of Machine Learning Research*, 12(10):2825–2830, 2011.
95. Mohamed El Kourdi, Amine Bensaid, and Tajje Eddine Rachidi. Automatic arabic document categorization based on the naïve bayes algorithm. In *The Workshop on Computational Approaches To Arabic Script-Based Languages*, pages 51–58, 2004.
96. Mostafa M Syiam, Zaki T Fayed, and Mena B Habib. An intelligent system for Arabic text categorization. *International Journal of Intelligent Computing and Information Sciences*, 6(1):1–19, 2006.
97. Xiang Zhang, Junbo Zhao, and Yann Lecun. Character-level convolutional networks for text classification. pages 649–657, 2015.

Chapter 7
Conclusions and Future Challenges

7.1 Conclusions

This book mainly discusses the architectures of intelligent networks as well as the possible techniques and challenges. As shown in Fig. 7.1, we first introduced the concept of NetworkAI, which is a novel paradigm that applying machine learning to automatically control a network. NetworkAI employs reinforcement learning and incorporates network monitoring technologies such as the in-band network telemetry to dynamically generate control policies and produces a near-optimal decision. We employ the SDN and INT to implement a network state upload link and a decision download link to accomplish a closed-loop control of a network and build a centralized intelligent agent aiming at learning the policy by interaction with a whole network.

Then, we discuss the possible machine learning methods for network awareness. With the rapid development of compelling application scenarios of the networks, such as 4K/8K, IoT, it becomes substantially important to strengthen the management of data traffic in networks. As a critical part of massive data analysis, traffic awareness plays an important role in ensuring network security and defending traffic attacks. Moreover, the classification of different traffic can help to improve their work efficiency and quality of service (QoS).

Furthermore, we discuss how machine learning can achieve network automatically control. One of the most critical issues in the network is finding a near-optimal control strategy. Examples include routing decision, load balancing, QoS-enable load scheduling, and so on. However, the majority solutions of these problems are largely relying on a manual process.

Therefore, to address this issue, in Chap. 4, we apply several artificial intelligence approaches for self-learning control strategies in networks. In addition, resource management problems are ubiquitous in the networking field, such as job scheduling, bitrate adaptation in video streaming and virtual machine placement in cloud

H. Yao et al., *Developing Networks using Artificial Intelligence*,
Wireless Networks, https://doi.org/10.1007/978-3-030-15028-0_7

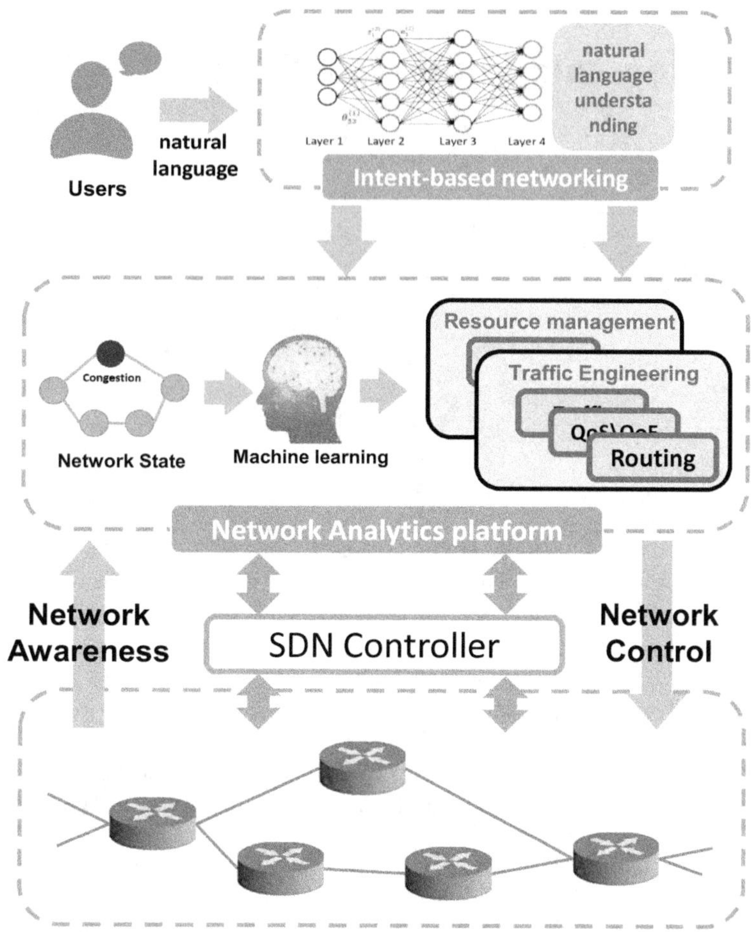

Fig. 7.1 NetworkAI conception

computing. In Chap. 5, we focus on how machine learning can optimal resource management decision.

Finally, we discuss the intent-based networking which allows a user or administrator to send a simple request—using natural language—to plan, design and implement/operate the physical network, so that it can reduce network complexity and improves automation by eliminating manual configurations. For example, an IT administrator can request improved voice quality for its voice-over-IP application, and the network can respond. For intent-based networking, the translation and validation system take a higher-level business policy (what) as input from end users and converts it to the necessary network configuration (how) by natural language understanding technology. To achieve intent-based networking, in Chap. 6, we focus on how machine learning can technology can be used in the natural language understanding in translation and validation system.

7.2 Future Challenges

7.2.1 New Hardware Architecture

Every innovation of the upper-level services is based on the extra improvement in the underlying hardware, for example, the Central Processing Unit (CPU) for general purpose computer, Digital Signal Processing (DSP) for the communication system, and Graphics Processing Unit for image processing. Similarly, for meeting the AI-driven networking age, the requirement for the specific AI networking processor is obviously.

Nowadays, networks generate millions of different types of flows every millisecond. Running AI agent in such massive volumes data is extremely challenging. The currently routers' computing power is far from being satisfied the requirement of AI&ML deployment. Especially, machine learning is a high computational intensive process. Recently, as the highly parallel, multi-core, multi-threaded processor, the GPU/TPU chip become the cornerstone for the age of AI. And some research works have shown that the GPU can improve the packet processing capability. However, due to the high-speed mass data processing (more than 10 Gb/s) and the demanding respond delay (less than 1 ms) in the future network, there is still a big gap between universal AI processing chip and actual deployment in the networking field.

7.2.2 Advancing Software System

Currently, the network data is becoming a big data challenges, e.g. threefold increase in total IP Traffic, >60% increase in devices and connections, telemetry data streamed in near real-time. Meanwhile, the geo-distributed characteristics of networking further impose more difficulties for the widespread deployment of network data analytic platform. For example, how to aggregates data like logs, metrics, and network telemetry; how to scales up to consume millions of flows per second; how to efficiently share knowledge among distributed network nodes. The currently end-to-end solution, which combines multiple technologies such as Apache Spark and Hadoop MapReduce, can be extremely complex and time-consuming. Therefore, a powerful scalable big data&AI analytic platform for networks and services is needed.

In addition, the software library for networking machine learning tasks is another enabler for AI-based networking. Machine learning framework offers a high-level programming interface for designing, training and validating machine learning.

However, current machine learning frameworks, such as Tensor-Flow, Caffe, and Theano, are designed for general-purpose, which is too overhead for networking domain. It needs further optimize to satisfy the requirement of the networking field, such as processing speed, low complexity, and light-weight.

7.2.3 *Self-Adaptive Network Protocol*

In networks, for exchanging various messages, the network protocols define rules, procedures, and formats for communication among network nodes. The current network protocols largely rely on the human being defined. To improve network flexibility and efficiency, there is still enough room for redesigning the network protocol to accelerate the efficiency of exchanging messages mechanism. Rather than matching patterns in a huge corpus of text, recently OpenAI released initial result that they trained the AI agents to invent a new language which is grounded and compositional. The multiple agents can automatically learn the communication protocol to coordinate their act enabled by machine learning. While these results are still far from the success of network protocol self-learning, it gives us a possibility of future network protocol evolution tendency.

7.2.4 *Promoting Machine Learning Algorithms*

While the machine learning algorithms are like a hundred flowers in bloom, current machine learning algorithms are driven by the existing business application, such as computer vision and natural language understanding. For example, convolutional neural networks are fascinating and powerful in image and audio recognition and even achieve superhuman performance in many tasks. However, the network presents a totally different theoretical mathematic model compared to the vision/NLP fields. The convolutional layers or recurrent layer may not work effectively in the networking domain. In addition, the network has massively more data and demanding response time which poses great challenges for machine learning deployment. Therefore, the demanding requirement and specific characteristics of networking require both adapting existing algorithms and developing new ones. The networking as a new application for machine learning will push forward of both machine learning and networking domain to a new stage.

CPSIA information can be obtained
at www.ICGtesting.com
Printed in the USA
LVHW081724020519
616424LV00002B/2/P

9 783030 150273